HIGH LATITUDE CLIMATE AND REMOTE SENSING

WILEY-PRAXIS SERIES IN REMOTE SENSING
Series Editor: David Sloggett, M.Sc., Ph.D.
Editor, *EARSeL Newsletter*

Co-Director, Dundee Centre for Coastal Zone Research, UK

Deputy Chairman, Cray Systems, UK

This series aims to bring together some of the world's leading researchers working in the forefront of the analysis and application of remotely sensed data and the infrastructure required to utilize the data on an operational basis. A key theme of the series is monitoring the environment and the development of sustainable practices for its exploitation.

The series makes an important contribution to existing literature encompassing areas such as: theoretical research; data analysis; the infrastructure required to exploit the data; and the application of data derived from satellites, aircraft and *in situ* observations. The series specifically emphasizes research into the interaction of elements of the global ecosystem publishing high-quality material at the forefront of existing knowledge. It also provides unique insights into examples where remotely sensed data is combined with Geographic Information Systems and high-fidelity models of the physical, chemical and biological processes at the heart of our environment to provide operational applications of the remotely sensed data.

Aimed at a wide readership, the books will appeal to professional researchers working in the field of remote sensing, potential users of the information and data derived from the application of remote sensing techniques, and postgraduate and undergraduate students working in the field.

EARTHWATCH: The Climate from Space

John E. Harries, Professor of Earth Observation, Imperial College, London, UK

GLOBAL CHANGE AND REMOTE SENSING

Kirill Ya. Kondratyev, Scientific Research Centre for Ecological Safety, Russian Academy of Sciences, St Petersburg, Russia; A. A. Buznikov, Electrotechnical University of St Petersburg, Russia; O.M. Pokrovsky, Main Geophysical Observatory, St Petersburg, Russia

HIGH LATITUDE CLIMATE AND REMOTE SENSING

Kirill Ya. Kondratyev, Scientific Research Centre for Ecological Safety, Russian Academy of Sciences, St Petersburg, Russia; O.M. Johannessen, Nansen Environment and Remote Sensing Centre, Bergen, Norway; V.V. Melentyev, Nansen International Environmental Remote Sensing Centre, St Petersburg, Russia

REMOTE SENSING AND GEOGRAPHIC INFORMATION SYSTEMS: Geological Mapping, Mineral Exploration and Mining

Christopher A. Legg, United Kingdom Overseas Development Administration, Forest and Land Use Mapping Project, Forest Department, Colombo, Sri Lanka

SATELLITE OCEANOGRAPHY: An Introduction for Oceanographers and Remote-sensing Scientists

Ian S. Robinson, Department of Oceanography, University of Southampton, UK

GEOGRAPHIC INFORMATION FROM SPACE: Processes and Applications of Geocoded Satellite Images

Jonathan Williams, Consultant, Space Division Logica plc, Leatherhead, UK

HIGH LATITUDE CLIMATE AND REMOTE SENSING

K. Ya. Kondratyev
Scientific Research Centre for Ecological Safety
St Petersburg's Russian Academy of Sciences, Russia

O. M. Johannessen
Nansen Environment and Remote Sensing Centre, Bergen, Norway

V. V. Melentyev
Nansen International Environmental Remote Sensing Centre, St Petersburg, Russia

JOHN WILEY & SONS
Chichester • New York • Brisbane • Toronto • Singapore

Published in association with
PRAXIS PUBLISHING
Chichester

Copyright © 1996 Praxis Publishing Ltd
The White House,
Eastergate, Chichester,
West Sussex, PO20 6UR, England

Published in 1996 by
John Wiley & Sons Ltd
in association with Praxis Publishing Ltd

Wiley Editorial Offices

John Wiley & Sons Ltd, Baffins Lane,
Chichester, West Sussex PO19 1UD, England

John Wiley & Sons, Inc., 605 Third Avenue,
New York, NY 10158-0012, USA

Jacaranda Wiley Ltd, G.P.O. Box 859, Brisbane
Queensland 4001, Australia

John Wiley & Sons (Canada) Ltd, 22 Worcester Road,
Rexdale, Ontario M9W 1L1, Canada

John Wiley & Sons (SEA) Pte Ltd, 37 Jalan Pemimpin 05-04,
Block B, Union Industrial Building, Singapore 2057

A catalogue record for this book is available from the British Library

ISBN 0-471-96093-4

Printed and bound in Great Britain by Hartnolls Ltd, Bodmin

Preface

A very popular cliché during the 1930s was that the Arctic is the kitchen of global weather and climate, although even at that time it was clear enough that such a judgement was based on an exaggeration. At the present-day stage of the development of global-scale three-dimensional climate modelling, the nature of this exaggeration is obvious. It is equally clear, however, that high latitude environmental processes deserve special attention in the context of global climate change and global change in general. The principal purpose of this book is to comment on such a conclusion on the basis of a discussion of specific features of high latitude environmental dynamics.

The aim of the first part of the monograph is to substantiate the scientific rationale for high latitude climate studies (Chapter 1) and to consider in more detail the subject of high latitude processes and global climate change. Both the similarities and differences between the two polar regions have been analysed with the use of observation data and numerical modelling results. Emphasis has been laid on the specific features of high latitude environmental conditions, such as the specific annual course of insolation, the presence of sea ice cover and the role of leads and polynyas, the special role of extended cloudiness, arctic haze and the antarctic ozone hole, and the high ecological sensitivity of polar regions, but hemispheric asymmetry with regard to anthropogenic climate signal due to greenhouse gases increases (an impact of tremendous thermal inertia of the southern hemisphere oceanic high latitude environment), etc.

Because of difficulties for *in situ* observations under high latitude conditions, satellite observations (remote sensing techniques) are an important tool, especially in the micro-wave region, since the application of remote sensing techniques in the visible and infra-red wavelength regions is aggravated by specific polar illumination conditions and clouds. The purpose of the second part of the book, devoted to a discussion of remote sensing techniques, is therefore limited mainly to a consideration of passive and active (radar) satellite remote sensing, with only a few comments concerning the potential of visible and infrared wavelength regions (Chapters 3 to 7). Emphasis has been laid on both the theoretical basis, and practical applications of microwave sensing.

The authors express their gratitude to Dr M. Miles for his valuable assistance in the preparation of the manuscript of this book.

The colour plate section is between pages 152 and 153.

Table of contents

1

Introduction: scientific rationale for high latitude climate studies

For some time studies of the climate-forming processes in the Arctic have been the subject of detailed scientific studies. Researchers highlighted the unique environmental circumstances in the Arctic—with the isolation of the ocean from the atmosphere by the highly variable and seasonal ice cover (Abelson, 1989; Arctic Climate..., 1994; Arctic Research, 1989; Arctic System..., 1990; Barry et al., 1993; Kondratyev, et al., 1990; Bjork, 1992; Climate Change..., 1990; Dynamics..., 1990; Federal..., 1985; Goody, 1980; Kondratyev, a, b, 1994, 1995, 1986, 1990, 1991a, b, 1992b; Nagurny, 1986; Nansen..., 1992; National Issues..., 1985; Report..., 1982, 1992, 1994; Scientific Concept, 1992; Sea Ice and Climate, 1988, 1990, 1992; Sea Ice Numerical, 1990; The Arctic Ocean..., 1994; The Ocean..., 1991; Untersteiner, 1982, 1984; Walsh and Crane, 1992; Witness..., 1994).

As has been emphasized (Polar regions..., 1993), polar regions are the sites of major interaction between the atmosphere, ice, oceans, biota and, in the Arctic, the land surface. These interactions influence the total Earth system through feedback, biogeochemical cycling, deep ocean circulation, atmosphere transport and changes in ice mass balance.

The report of the second session of the World Climate Research Programme (WCRP) steering group on global climate modelling (Global Climate..., 1992) and the documents of the WCRP Joint Scientific Committee (JSC) emphasized that information has been accumulated on the effect of the Arctic Ocean on the global climate system, though, so far, different physical processes, determining the ocean–atmosphere–ice cover interaction remain poorly studied. In this connection, further realization of the focused programmes of field experiments and monitoring is necessary, as well as numerical modelling, to study the processes of ice formation in the Arctic Ocean, of stratification and circulation of water masses, advection of energy and fresh water, ice cover dynamics, extended cloudiness-radiation interaction, and global warming impact (Kondratyev and Kotliakov, 1991; Kondratyev, 1992a; Kotliakov and Krenke, 1982; Lappo et al., 1990; Marchuk, et al., 1986, 1990; Matveev, 1991; Mysak et al., 1989, 1990, 1991, 1992; Namias, 1981; Newell and Wu, 1994; Newell and Zhu, 1994; Rango and Martinec, 1994; Raschke and Jacob, 1983; Romanov et al., 1987; Royer et al., 1990; Shapiro, 1991). For these reasons a number of programmes such as Arctic System Science: A Plan for Integration (Witness..., 1994); Man's Future..., 1990; Arctic Climate System Study (1994); Nansen Centennial Arctic Programme (1993) were launched.

1.1 PROCESSES IN THE ARCTIC

Aagard (Arctic Climate..., 1994) has pointed out that climate models generally suggest that global climate change is greatly amplified in the Arctic. Conversely, it is increasingly believed that the Arctic plays a significant role in global climate control, chiefly through its influence on the large-scale thermohaline circulation of the oceans (the "conveyor belt") and on the global heat budget (e.g., the snow and ice albedo feedback). However, neither the observational base, or our knowledge of the controlling physical processes, or our ability to model the role of the Arctic in global climate are at present sufficient to realistically assess the principal issues linking the Arctic with global climate. This is why the Arctic Climate System Study (ACSYS) project was developed in 1991 as the WCRP initiative and a practicable programme for the next decade to assess the role of the Arctic in global climate (Arctic System..., 1990; Scientific Concept..., 1992; Kondratyev and Kotliakov, 1991). Five areas are emphasized: (1) ocean circulation (see also Nansen Centennial..., 1992; The Arctic Ocean..., 1994); (2) sea ice climatology (see also Sea-Ice and Climate..., 1988, 1990, 1992; Sea Ice Numerical, 1990); (3) the Arctic atmosphere; (4) the hydrological cycle; and (5) modelling. The scientific goals of ACSYS, which started its main observational phase on January 1994 and will continue for a ten-year period includes the three main objectives (Arctic Climate..., 1994):

(1)   understanding the interaction between the Arctic Ocean circulation, ice cover and the hydrological cycle;
(2)   initiating long-term climate research and monitoring programmes for the Arctic;
(3)   providing a scientific basis for an accurate representation of Arctic processes in global climate models.

The Arctic Ocean Circulation Programme of ACSYS consists of the four components:

(1)   the Arctic Ocean Hydrographic Survey to collect a high-quality, hydrographic database representative of the Arctic Ocean;
(2)   the Arctic Ocean Shelf Studies which are aimed at understanding how the shelf processes partition salt- and fresh-water components; at defining the dynamics and thermodynamics of the shelf waters as well as other processes;
(3)   the Arctic Ocean Variability Project designed to assess the variability of the circulation and density structure of the Arctic Ocean;
(4)   the Historical Arctic Ocean Climate Database Project aimed to establish a universally available digital hydrographic database for the Arctic Ocean for analysis of climate-related processes and variability, and to provide a data set suitable for initialization and verification of arctic climate and circulation models.

The ACSYS sea ice programme includes three main components:

(1)   establishing an Arctic Basin-wide sea ice climatology database;

(2)   monitoring the export of sea ice through the Fram Strait;

(3)   arctic sea ice process studies.

One of the main tasks of the ACSYS arctic sea ice programme is to establish a climatology of ice thickness and ice velocity. Such data will be supplied by the WCRP Arctic Ice Thickness Project, the International Arctic Buoy Programme, sonar profiling naval submarines and unmanned vehicles, airborne oceanographic lidars, and polar satellites carrying appropriate instruments.

The arctic atmosphere provides the dynamic and thermodynamic forcing of the Arctic Ocean circulation and sea ice. Key directions of research include problems such as: cloud–radiation interaction, air–sea interaction in the presence of ice cover (impacts of polynyas and leads are of special interest), arctic haze etc.

Primary ACSYS efforts within the project of the Hydrobiological Cycles in the arctic region are aimed at:

(1)   the documentation and intercomparison of solid precipitation measurement procedures used in high latitudes and

(2)   the development of methodologies for determining areal (regional) distributions of precipitation from station data.

There are two relevant data-archiving efforts: Arctic Precipitation Data Archive (APDA) and Arctic Run-off Data Base (ARDB).

The principal purpose of the ACSYS modelling programme is simulations of climate variations in polar regions which arise from the interaction between atmosphere, sea ice and ocean.

A much broader Arctic System Science: A Plan for Integration (ARCSS) was published in September 1993 (Witness..., 1994). This programme was initiated to gain a better understanding of interaction within the arctic system, specifically how it responds to global change and how it interacts with the rest of the Earth's system. Integration and synthesis are the highest general priorities of ARCSS, which consists of three linked components: Palaeoenvironmental Studies; Contemporary and Process studies; and Synthesis, Integration and Modelling Studies. Palaeoenvironmental Studies include the Greenland Ice Sheet Project Two (GISP2), which will be completed in 1996, and the ongoing Palaeoclimates of Arctic Lakes and Estuaries (PALE). Contemporary and Process studies include Ocean/Atmosphere/Ice Interactions (OAII) and Land/Atmosphere/Ice Interactions (LAII).

LAII research has three main goals (Witness..., 1994):

(1)   to estimate important fluxes in the region, including the amount of carbon dioxide, and methane reaching the atmosphere, the amount of river water reaching the Arctic Ocean, and the radiative flux back to the atmosphere;

(2)   to predict how possible changes in the arctic energy balance, temperature and precipitation will lead to feedback affecting large areas; this incorporates changes in water budget, duration of snow cover, extent of permafrost, and soil warming, wetting, and drying; and

(3)   to predict how the land and fresh-water biotic communities of the Arctic will change, and how this change will affect future ecosystem structure and function.

A major LAII research project is the Flux Study; its principal purpose is a regional estimate of the present and future movement of materials between the land, atmosphere, and ocean in the Kuparuk river basin in northern Alaska.

Of the nineteen LAII projects three are part of the International Tundra Experiment (ITEX) which looks at the response of plant communities to climate change. Three others are concerned with atmosphere processes, including weather pattern affecting snowmelt, arctic-wide temperature trends, and water vapour over the Arctic and its relationship to the atmospheric circulation and surface conditions. One project deals with response of large birds to climate and sea-level change at river deltas, and one studies the balance and recent volume changes of McCall Glacier in the Brooks Range.

Synthesis, integration and modelling studies are intended to foster linkages and system-level understanding. Research on the past contemporary relationship of humans to global climate change is thought to be critical to understanding the consequences of global change in the Arctic.

There are a number of ARCSS data projects, including: LAII Flux Study Alaska North Slope data sampler CD-ROM; OAII Northeast Water (NEW) polynya project CD-ROM; Arctic solar and terrestrial radiation CD-ROM etc.

A list of the OAII components includes the joint US/Japan cruise, the Western Arctic Mooring project, and the Northeast Water Polynya project. Among other OAII projects the most notable are the US/Canada Arctic Ocean Section and the Surface Heat Budget of the Arctic Ocean (SHEBA) project.

An outstanding effort has been accomplished in 1994 within the Canada/US 1994 Arctic Ocean Section when the two icebreakers entered the ice in the northern Chukchi Sea on 26 July 1994, reached the North Pole on 22 August, and left the ice northwest of Spitsbergen on 30 August, thereby completing the first crossing of the Arctic Ocean by surface vessels. This voyage will greatly alter our understanding of biological productivity, the food web, ocean circulation and thermal structure, and the role of clouds in the summer radiation balance, as well as the extent of contamination and spreading pathways (especially related to radionuclides and chlorinated organics), and the extent and effects of sediment transport by sea ice (Witness…, 1994).

In connection with the Sheba project the US Department of Energy's Atmosphere Radiation Measurement (ARM) programme indicated its intention to develop a Cloud and Radiation Testbed (CART) facility on the North Slope of Alaska. The principal focus of this programme will be on atmospheric radiative transport, especially as modified by clouds (which impacts the growth and decay of sea ice), as well as testing, validation, and comparison of radiation transfer models in both the ice pack and arctic coastal environment.

The recent meeting of the International Arctic Science Committee (IASC) has identified the following four science priorities (Witness…, 1994):

(1)   arctic processes relevant to global systems;
(2)   effects of global change on the Arctic and its peoples;
(3)   natural processes within the Arctic; and
(4)   sustainable development in the Arctic.

The following areas in arctic global change research have been considered the most significant:

(1)   terrestrial ecosystem;
(2)   mass balance of glaciers and ice sheets;
(3)   regional cumulative impacts; and
(4)   human dimensions.

An important aspect of studying high latitude environmental dynamics are assessment of the impact of potential anthropogenic climate warming. In this context Frederick (1994) has formulated key issues of integrated assessments of the impact of climate change on natural resources. Specific project objectives include: (1) characterizing the current state of natural science and socioeconomic modelling of the impacts of climate change and current climate variability on forests, grasslands, and water; (2) identifying what can be done currently with impact assessments and how to undertake such assessments; (3) identifying impediments to linking biophysical and socioeconomic models into integrated assessments for policy purposes; (4) recommending research activities that will improve the state of the art and remove impediments to model integration.

The following questions are supposed to be answered:

— How will the overall system (physical–biological–economic) respond to various imposed stressors?
— How do the uncertainties in the component models add up to give an overall system response uncertainty?
— Is society made more vulnerable to extreme natural events either by changing those events or by reducing human ability to respond with corrective action?
— How likely is it that the consequences of climate change will be severe or catastrophic?
— What is at risk and when is it a risk?
— What are the likely impacts on the landscape and the hydrological system?
— How might the boundary conditions and the overall productivity of the forests, grasslands, and other rangelands be affected?
— How might increasing carbon dioxide levels affect crops and food supplies for humans, livestock, and wildlife?
— What are the socioeconomic consequences of these physical and biological changes?
— What are the likely consequences for ecosystems of mitigation actions?
— Can the costs associated with climate change be reduced through natural adaptation of ecosystems or policy-initiated adaptation?

Frederick (1994) has emphasized that the accumulated results of many regional and local climate impact assessments may help provide informed answers to these questions. Nevertheless, the uncertainties surrounding both the nature and the impacts of any future climate change are likely to remain very large, precluding precise estimates of the net benefits associated with alternative policy responses. Even if the range of uncertainty were diminished, it might still be difficult to justify specific measures on narrow economic grounds because (as noted above) the impacts on natural resource systems, are apt to be poorly reflected in standard benefit–cost analysis.

Mendelsohn and Rosenberg (1994) have formulated the following questions relevant to global warming effects in the area of ecological and water resources:

— Do changes in ecosystems provide important feedbacks to the natural carbon, nitrogen, and methane cycles? For example, will the natural sinks or emitters be affected by changing precipitation, temperature, and CO_2 levels?
— What are the appropriate output measures of ecosystem component models? What are the ecological effects of climate change that policy analysts use to determinate the importance of an ecosystem change?
— What climate change-driven shifts in ecosystem boundaries can be predicted?
— Will these effects be subtle and small or large and dramatic and over what time frame and spatial dimensions?
— Will climate change cause a change in the productivity of valuable market or nonmarket species? For example, to what extent will some forests grow more quickly or more slowly. Will desired nonmarket species such as bear, elk, and bald eagles be more or less plentiful?
— What species could be lost with rapid climate changes? How do the vulnerable species break down by type and geographic distribution? How should conservation policies adapt to a world requiring change?
— How are ecosystems likely to change as the climate evolves over time: will there be a large increase in early succession species and where?
— How will average flows in rivers change with greenhouse warming? How will these flows change over seasons? Will the probabilities of catastrophic events change?
— What values do people assign to the changes in ecosystems caused by climate change? Which changes are important and which are minor? Can a value be assigned to non-use?
— How much should society be willing to pay to reduce the probability of losing specific species? If different scenarios favour different species, how should society trade between these outcomes?
— What impact do ecosystem changes have upon the economy? For example, how will climate change affect grazing, commercial fishing, timber, or commercial tourism?

It has been suggested (International..., 1994) that priority programme areas and relevant projects are as follows:

(a)   Impacts of global changes on the arctic region and its peoples
        — Regional cumulative impacts
        — Effects of increased UV radiation.
(b)   Arctic processes of relevance to global systems
        — Mass balance of glaciers and ice sheets
        — Terrestrial ecosystems and feedback on climate change
(c)   Natural processes within the Arctic
        — Arctic marine/coastal/riverine systems
        — Disturbance and recovery of terrestrial ecosystems
(d)   Sustainable development in the Arctic
        — Sustainable use of living resources

— Dynamics of arctic populations and ecosystems
— Environmental and social impacts of industrial development.

A special place belongs to a research strategy for the 1990s onward studying the dynamics of the Arctic Ocean which has been developed by the European Committee on Ocean and Polar Science (ECOPS) (Dynamics..., 1990; The Ocean..., 1991; Kondratyev and Kotliakov, 1993; Kondratyev, 1995; Lange, 1993). The Dynamics of the Arctic Ocean Programme includes such parts as: (1) sea ice; (2) ocean circulation; (3) exchange processes; (4) biogeochemical processes.

1.2 PROCESSES IN THE ANTARCTIC

There are a number of similarities in processes in both polar regions. As far as problems of climate change are concerned the following priorities deserve special attention (Arctic System..., 1990):

— extremal annual variations of radiation, extent of snow and sea ice cover as well as energy exchange between the surface (land and ocean) and the atmosphere;
— the presence of large water reservoirs in ice sheet of the Antarctic and the Arctic (in Greenland);
— the existence of ocean water masses with weakly stratified density profile which favours vertical mixing reaching great depths;
— sea water phase change when ice forms or melts.

There are, however, significant differences between processes in the Arctic and the Antarctic due to radically different distributions of ocean and land as well as other factors (Budd, 1991; Harris and Stonehouse, 1991; Hibler, 1992; Hibler and Thorndike, 1992; Lubin, 1994; Mancini et al., 1992; Moore et al., 1991; Peng et al., 1992; Radiation and Climate, 1992; Radionov, 1994; Radionov et al., 1994; Raschke et al., 1992). An important problem is the role of Antarctica and the Southern Ocean in the global climate change (Genthon, 1994; Hempel, 1994; Houghton, 1984; Katsov et al., 1993; Kondratyev, 1988, 1992a; Oerlemans and van der Veen, 1984; Peltier, 1993; Pierce et al., 1994; Stössel, 1992; Tzeng et al., 1993; Weatherly et al., 1991; Weller, 1992a, b, 1993; Wilkness, 1989).

Weller (1992b) has pointed out that Antarctica plays a critically important role in global environmental change owing to interaction between the atmosphere, ice cover, ocean and oceanic biota which are accompanied by various feedbacks with the participation of such processes as biogeochemical cycles, circulation in the deep ocean, transport of energy and pollutants in the atmosphere, changes of mass balance and so on. On the other hand, Antarctica is a region of high sensitivity to global change impacts, which makes this region important from the viewpoint of early detection of global change and prompts detailed environmental monitoring in Antarctica as an urgent necessity.

A very important specific feature of environmental dynamics in Antarctica is the formation of the ozone hole (Adriani et al., 1992; Bojkov, 1993; Deshler et al., 1992; Gobby and Adriani, 1993; Hofmann et al., 1991, 1992; Massie et al., 1984; Park and Russel, 1994; Pierce et al., 1994; Rosen et al., 1992; Taylor et al., 1994; Kondratyev,

1989; Russel et al., 1993). It has become clear that the following conditions are necessary for the development of an ozone hole with the rapid drop (1–2% per day) of the total ozone content (TOC):

— low stratospheric temperature which is required for the formation of polar stratospheric clouds (PSC);
— presence of aerosols or PSC particles whose surfaces serve as places of catalytic ozone destruction owing to relevant heterogeneous chemical reactions;
— increased chlorine concentration (chlorine is being produced through dissociation of chlorofluorocarbons (CFCs) transported to the stratosphere);
— stable circumpolar vortex (CPV) which creates a closed space for heterogeneous chemical reactions.

Cattle et al. (1992) have studied the response of the Antarctic climate in general circulation interactive model experiments for the atmosphere–ocean–land surface system with transiently increasing (1% per year) carbon dioxide concentration. They show the existence of strong climate asymmetry between the hemispheres. Owing to the oceanic impact in the southern hemisphere greenhouse climate warming above the sea ice cover around Antarctica is very small (see also Huybrechts, 1990; Huybrechts and Oerlemans, 1990).

Of great interest is the dynamics of the ice sheets of Greenland and the Antarctic (the glaciers of Greenland can be the most sensitive indicator of global climate changes (Glaciers..., 1985)).

Interesting estimates of the water vapour input to the Antarctic have been obtained by Koster and Svarez (1993). Based on the GISS AGCM calculations (over the 8° lat. by 10° long. grid), these authors calculated the contribution of evaporation from the surface of various oceans to the formation of precipitation in the Antarctic in July. It turned out that for each antarctic site considered the contribution to precipitation was determined by the impact of several SH distant regions of the World Ocean (11 regions have been selected by their characteristic SST). So, for example, 6% of precipitation in the Antarctic are determined by evaporation from the surface of the Pacific regions, where SST varies within 5–10°C. The total contribution of the Pacific, as well as Atlantic and Indian Oceans to the antarctic precipitation is, respectively, 36.7% and 56.9%, with these two macro-regions being the sources of water for various parts of the Antarctic. A negligible part of precipitation (about 0.6%) arises from local water sources.

The contribution to precipitation of water contained at the moment in the atmosphere is only 0.2%. Connolley and King (1993) have discussed atmospheric water vapour transport to Antarctica inferred from radiosonde data.

The problem of the gas-to-particle formation of condensation nuclei and their effect on cloud cover is full of uncertainties (Harte and Williams, 1988; Kvenevolden, 1988). Of great importance is the problem of the methane cycle in the Arctic. The problem of gas hydrates has also gained new importance (Cox et al., 1993; Hines and Morrison, 1992; Loehle, 1993; Morrissey and Livingstone, 1992).

The problem of the strong enhancement of energy- and mass-exchange in the polynyas and leads areas which is important for both polar regions is still urgent (Meleshko, 1991; Meleshko et al., 1991; Nagurny, 1986; Petropavlovskaya, 1988; Programme, 1989;

Romanov et al., 1987; Dey, 1981; Dey et al., 1979; Dey and Feldman, 1989; Treshnikov et al., 1991; Doskey and Gaffney, 1992; Glowienska-Hense and Hense, 1992; Fan et al., 1992; Harriss et al., 1992; Hibler and Thorndike, 1992; Kottmeier and Engelbart, 1992; Morison et al., 1992; Mysak and Huang, 1992; Rottier, 1992; Serreze et al., 1992a, b, c; Simmonds and Budd, 1991; Smith et al., 1990; Whiting et al., 1992; Report..., 1992). Special attention should be paid to the processes of formation of bottom waters in the polar regions and, in this connection, the role of the Arctic Ocean in the global carbon cycle.

In the context of the problem "Polar Regions and Climate" of great interest is the programme to study the climate of polar regions (Programme..., 1989) developed in the Institute of Arctic and Antarctic (IAA) as well as programme developments such as "Arctic Research for an Arctic Nation" (Arctic Research..., 1989) and "Nansen Centennial Arctic Programme" dedicated to the centennial anniversary of the expedition on "Fram" (Nansen Centennial..., 1992).

1.3 SPECIFIC FEATURES OF CLIMATE FORMATION IN POLAR REGIONS. REQUIREMENTS FOR OBSERVATIONS AND NUMERICAL MODELLING

Baker (1981) considered principal manifestations of the climatic effect of the processes in the polar oceans. In conditions of the Arctic, these processes include: (i) the effect of heat transport in the ocean across the Greenland–Spitsbergen passage on the variability of pack ice; (ii) possible effect of the diversion of part of the Siberian rivers run-off on the decrease of the fresh water influx to the Arctic basin and subsequent transformation of sea water stratification; (iii) the impact of the ocean and pack ice on the formation and variability of the summertime overcast stratus clouds; (iv) intensification of climate warming in high latitudes due to increasing CO_2 concentrations and its possible consequences (more intensive CO_2 release by the ocean, the effect on ice melting, etc.); (v) cyclogenesis near the edge of the polar ice cover; (vi) the influence of the atmosphere–ocean interaction in the Greenland–Norwegian Seas on the formation and variability of the North Atlantic deep waters (the conditions of balances between the output and input of these waters along the eastern coastline of the Atlantic; the impact of the Gulf Stream on Europe).

Semtner (IUGG..., 1991), characterizing the role of polar oceans in the formation of climate, pointed out that the polar oceans receive the heat from the mid-latitude oceans (mainly through advection in the Arctic and eddy transport in the Antarctic). This heat affects strongly the annual course and the interannual variability of ice conditions in both hemispheres. On the other hand, the sea ice cover dynamics affects substantially the radiation budget and meteorological conditions near the ocean's surface.

Of importance is the effect of polar oceans on the deep-water circulation of the world ocean. Most of the heat from the surface layer of the ocean in the Arctic and Antarctic is spent on the formation of thermohaline circulation. The complex totality of the processes of convection in the middle layers of the ocean, in the shelf zones, etc., results in the formation of bottom waters in the North Atlantic. Similar processes take place in the southern hemisphere. An adequate account of the processes in the polar oceans is of principal importance for the solution of the problem of future climate predictions and explanation of palaeoclimatic variations. It is important here, as Cattle (IUGG..., 1991) notes, to take into consideration special processes taking place in the Arctic and Antarctic.

Mysak (Mysak et al., 1990; Mysak and Power, 1991; Mysak and Huang 1992) has laid special emphasis on the role of the interdecadal climatic variability in the Arctic that manifests itself in the form of SST variations, salinity and ice cover, in the formation of global climate. An analysis of observational data revealed the processes of interaction in the atmosphere–ice cover–ocean system, determining an appearance of negative feedback responsible for the shelf-maintaining interdecadal climatic variations. A characteristic indicator of such variations in the Greenland Sea is the presence of intermittent regimes of extended ice cover over the layer of cool fresh waters, which determines the winter-time suppression of deepwater convection, and a small area of the ice cover over the waters of stronger salinity, which stimulates a convection. Such a cycle brings forth decadal variations in the rate of deepwater formation in the North Atlantic and, as a result, the interdecadal variations of the global climate due to the effect of thermohaline circulation. It is important to point out that the problem of interdecadal climate variation has recently attracted great attention, especially in the context of an early detection of the anthropogenic greenhouse climate change signals.

The unsolved problems of the Antarctic climate are as follows: (i) heat balance of the continent and adjacent seas (it remains unclear whether the heat flux can be directed northward from the Atlantic countercurrent in the Southern Atlantic and southward in the Pacific Ocean); (ii) the role of the meridional heat transport as a possible factor that determines the inadequacy of the classical conception of the oceanic circulation and effects the climate (here of importance is the interconnection of heat transport and SST anomalies determined by the prevailing role of strong variations in the extent and concentration of the polar ice cover with its high albedo and heat-insulating effect in the formation of heat balance and climate: in the period of maximum development of the antarctic ice the size of the continent practically doubles, while in the Arctic the amplitude of the interannual ice cover variability reaches 50% of the annual mean value); (iii) the connection of SST anomalies south of New Zealand with El Niño; (iv) the effect of the SH oceans on the poleward heat transport in the atmosphere; (v) cyclogenesis near the ice cover edge and near the oceanic polar front; (vi) fast "barotropic" response of the antarctic countercurrent to changes in the zonal wind field and its connection with climate; (vii) contribution of the 6-month oscillations in the atmosphere and ocean to climate variability. The factors of variability in the extent and concentration of the polar ice cover and in ice transport by sea currents have been studied inadequately, so far (in the Arctic, for example, the ice output from the Arctic Basin east of Greenland is as important from the viewpoint of the formation of the regional heat balance as the heat transported by the sea currents).

To solve the enumerated (and other) unsolved problems, first of all it is necessary to expand the programme of observations in high latitudes, especially long-term observations in the key regions and special problem-oriented complex programmes (atmosphere–ocean interaction at the polar ice edge, monitoring the properties of sea water under the ice cover and the ocean level, meridional heat transport, etc.). For this purpose increased effort is needed to improve the observational means available and to develop new ones (especially radio-altimetry and acoustic tomography).

Study of the climatic effect of the cryospheric properties (sea ice and snow cover) and processes taking place there is one of the WCRP first priorities (Abelson, 1989; Alekseev et al., 1991; Alekseev, 1994; Barnett et al., 1989; Barry et al., 1987, 1993a, b; Bromwich and Robasky, 1993; Bromwich et al., 1994; Browell et al., 1992; Charles et al., 1994; Cheng and Preller, 1992; Climate Change, 1990; Cohen and Rind, 1991; Cohen, 1994; Covey et al., 1991; Gleick and Rango, 1994; Goody, 1980; Goody and Yung, 1989; Hakkinen, 1993; Hakkinen and Mellor, 1992; Hibler, 1992; Houghton, 1984; Johannesen et al., 1992, 1994, 1995; Kondratyev, 1986, 1988, 1992a, 1995; Kondratyev and Johannessen, 1993; Ledley, 1988, 1991, 1993; Legenkov, 1992; McPhee, 1992; Namias, 1981; Omstedt and Wettlaufer, 1992; Parkinson, 1991; Parkinson and Cavalieri, 1989; Peltier, 1993; Rasmussen et al., 1993; Report..., 1982; Role..., 1990; Sea Ice and Climate:,1988, 1990, 1992; Zakharov, 1981). The report of the WMO/CAS-JSC-CCCO meeting of experts held on 24–29 June 1982 contained the survey of the role of sea ice in climatic changes (Report..., 1982).

The most important problem is to develop the coupled climate models that would take into account the processes in the cryosphere, since: (i) the strong annual change and interannual variability of the ice cover are followed by respective variations in surface albedo as well as heat- and water-exchange between the ocean and the atmosphere; (ii) the ice cover dynamics intensifies the variability of the temperature regime, which is the result of a substantial positive albedo feedback in the global climate system; (iii) the ice cover is an important indicator of climatic variability (in particular, that of anthropogenic origin).

The following four directions of research are of primary importance: (i) the substantiation of more reliable techniques to parameterize the processes in the cryosphere (changes in albedo, heat- and moisture-exchange, consideration of the contribution of polynyas and leads; cloud-radiation interaction, etc.); (ii) study of interaction of the "ocean–ice cover–atmosphere" system at the ice cover edge and taking account of this interaction in climate theory; (iii) analysis of the sensitivity of climate models to realistic changes in the extent and structure of the ice cover; (iv) continuous monitoring of the ice cover characteristics from satellites, and especially at the ice cover edge. Apparently, of key importance is an account of sea ice dynamics in the Antarctic, characterized by the strongest seasonal variability.

Of particular interest are the following events in the Antarctic: (i) circulation in the Weddell Sea; (ii) sub-ice sea stratification; (iii) antarctic bottom waters, the main source of which is the Weddell Sea. The most important problem is the development of the coupled climate models which take into consideration (parameterize) the processes in the sea ice, based on the diagnostic analysis of the observational data.

Table 1.1 illustrates the requirements for observational data aimed at obtaining the data that characterize: (i) the annual change of the extent and concentration of the ice cover averaged over a large time period, as well as the interannual variability; (ii) drifting of the antarctic sea ice; (iii) the distribution of the ice cover thickness over the Antarctic; (iv) distribution and relative share of "cake ice"; (v) the field of sea currents; (vi) fields of temperature and pressure in the atmosphere; (vii) the annual change of the vertical heat

Table 1.1. Requirements for sea ice observation data for numerical climate modelling

Parameter	Error	Spatial resolution (km)	Temporal resolution
Ice cover edge	10 km	50–200	Averaged over 3 days, 1 week, and 1 month
Share of open water	5–10%	50–200	ditto
Type (thickness) of ice	5 levels[a]	50–200	ditto
Ice drift	10% drift for 1 month	50–200	ditto

[a] Open water; 0–10 cm; 10–70 cm; 70–200 cm; >200 cm.

flux. Of great importance is a realization of the problem-oriented field experiments to study the most important aspects of the effect of sea ice on climate. Such experiments should be planned based on the respective simulation numerical experiments.

As Allison (Report..., 1982) noted, the first approximate climate models had taken into account only the albedo feedback determined by the formation (or melting) of sea ice. In fact, the ice cover is a deforming barrier which reacts to the atmospheric forcing (and, in its turn, affects the atmosphere), the processes of interaction among the ocean, ice and atmosphere being very complicated. The ice cover variability determines an important contribution to the advective heat- and salt-exchange between polar regions and the rest of the globe.

Recently it has become clear that the ice cover should not be treated in climate theory as a fixed surface with varying albedo (depending on the presence of continuous ice cover, thawed patches, etc.). Besides, of critical importance are characteristics such as the thickness and concentration of ice cover, which control the motion of ice through its interaction with the atmosphere and ocean, and the formation and disintegration of ice cover.

The contrast geographical conditions of two polar regions determine their marked thermodynamic and dynamic special features. The antarctic sea ice is characterized by the seasonal variability, its thickness constitutes about 1.5 m, and the extent undergoes great annual changes (from 2.5×10^6 to 20×10^6 km^2), the external ice boundary is not confined by the continent (as it is in the Arctic), which determines the presence of a considerable meridional advection and free exchange with the ocean. The Antarctic Circumpolar Current (ACC), which ensures water transport of about 130×10^6 m^3 s^{-1}, is much more intensive compared to the current in the Arctic Ocean. The meridional heat transport across the ACC zone reaches 0.28×10^{15} W and is, apparently, determined by gyres and not a regular convection (as in the East Greenland current). The major ACC component—the gyre in the Weddell Sea—is responsible for, at least, 40% of total meridional heat transport in the SH ocean.

The multi-year ice about 4 m thick prevails in the Arctic, with a longitudinal asymmetry of distribution owing to specific contours of the continental coastline "compressing" the arctic ice. The latter determines a much smaller amplitude of the annual change of the ice cover extent (from 7×10^6 to 14×10^6 km^2), and the zone of the ice edge is not as affected by winds and currents as near the Antarctic.

Characteristic features of the ice cover and relevant processes in the Arctic and Antarctic determine the necessity of special studies of the role of ice in the formation of climate in both polar regions. The need for interactive climate models taking into account (parameterizing) the processes that determine the dynamics (growth, disintegration, splitting, melting) of sea ice is a longstanding problem; of particular importance is a description of the processes in the regions of polynyas and leads, as well as near the ice edge.

Estimates have shown that in the Arctic about half the heat release from the ocean to the atmosphere takes place in winter through polynyas, leads and ice less than 0.8 m thick, with a total area less than 10% of the whole ice cover extent. An adequate parameterization of the processes in the atmospheric boundary layer and the ocean is of critical importance. The turbulent heat exchange in the upper layer of the ocean is a major factor of the thermodynamics and the dynamics of the ice-covered ocean. In particular, the ocean is seriously affected by salt release resulting from ice formation and causing variations in the upper ocean layer stability and advection of salt added to the ocean in the regions of ice formation and removed in other regions, where the ice melts.

Since aircraft observations over the arctic leads and polynyas revealed the trails of heat rising up to 4 km (i.e. penetrating the layer of inversion, promoting, thereby, tropospheric heating), Glendening (1992) calculated the lead-to-atmosphere heat transport and its vertical distribution, bearing in mind an analysis of the necessary parameterization of such a phenomenon in large-scale atmospheric models (a characteristic width of leads varies within a range of 1–10 km). Since the upward heat flux over the leads and polynyas is, as a rule, two orders of magnitude greater than heat input through the ice cover, despite a relatively small area of leads and polynyas, their contribution to the heat input to the atmosphere can be greater than that through the ice cover. The estimates have shown that with the relative area of leads increasing from 1.1% to 4.3%, the mean annual, mean zonal SAT in the north polar region should rise by 1 K. The initial trail of the heat outbreak moves windward, which leads to the formation of a layer of the upward motions over the edge of leads (this is testified to by calculations of the 3-D structure of the trail). Part of the upward heat flux comes back to the ice cover surface in the lee zone, where the downward heat flux originates, an account of which is important to assess the contribution of leads to the warming of the atmosphere. The air temperature variations at a given level simulated with a large-scale climate model is determined by the value of horizontally averaged heat flux from the leads, which exponentially decreases with altitude, provided the wind does not flow along the leads (in the latter case the vertical advection is very important).

A major aspect of the problem of the impact of the sea ice on climate is the heat exchange at the "atmosphere–ice cover–ocean" system interface. The decisive effect on the annual change of the ocean upper layer temperature is made by the latent heat resulting from ice formation or melting, the buoyant flow connected with the salt exchange, and the atmosphere-to-ocean momentum transfer across the ice cover.

Apparently, an important factor of the ice cover dynamics in the Antarctic is a still poorly studied heat input from ocean, reaching 30 W m^{-2} (this can be a reason for the rapid disintegration of the ice cover in spring). In the arctic conditions, where the strong stability of stratification suppresses the vertical exchange, the role of the heat flux from the ocean is, probably, negligible.

The emphasis has been placed on a number of inadequately understood processes and phenomena that occur near the ice cover edge such as upwelling, convection, fronts, gyres and geostrophic flux along the edge. The construction of the coupled climate models requires a reliable substantiation of the parameterization of the ice cover dynamics with account of ice rheology. A substantial contribution to the heat- and salt-exchange in the ocean is made by the ice advection, particularly strong near the ice cover edge.

Pollard (Report..., 1982) emphasized that the use of AGCM to describe the climate of polar regions revealed considerable differences with the observational data, characterized by: (i) the SAT overestimated by < 20°C in both hemispheres, especially in winter; (ii) eddy heat flux towards the surface overestimated by < 50 W m^{-2}. In the models with prescribed ice cover the arctic ice is often too thin (< 1 m), and the extent of the antarctic ice is underestimated (there is no ice in summer, and in winter its thickness is < 50 cm).

Apparently, major reasons for the differences are: (i) overestimated summertime global radiation due to the underestimated cloud cover; (ii) overestimated downward sensible heat flux in winter; (iii) overestimated diffusion horizontal heat flux in the ocean and the underestimated Eckman transport to the equator across the 60°C parallel (the latter refers to the models which take into account the oceanic circulation).

Before introducing the improved techniques of parameterization, it is expedient to "adjust" the AGCM available, to reach an agreement with the observed extent and thickness of the ice cover. Apparently, the most serious defects of parameterization arise from neglecting the role of polynyas and leads in the Antarctic and the ice dynamics in the Arctic. To reveal the model components responsible for differences, a comparison with observational data on the surface heat balance is very important. From the viewpoint of the oceanic general circulation models (OGCM), a description of heat flux towards the lower boundary of the ice cover in the Antarctic controlled by the speed of upwelling and (or) the input of the warmer waters of halocline to the mixed layer, is very important. If these processes are determined by the penetrating eddy mixing, a model will be needed of the mixed layer of varying thickness with the "buoyant fluxes" depending on salinity, which is determined by the processes of melting and freezing. The problem of parameterization of the processes in the zones of leads and polynyas remains unsolved.

Simple thermodynamic models of a nonhomogeneous ice layer should be specified, bearing in mind: (i) the ice dynamics (for example, the output of ice from the Arctic across the Greenland Sea and the Eckman divergence round the Antarctic; (ii) the varying share of leads in each cell of the spatial grid.

The coupled AGC/OGC models can be tested by: (i) testing the reliability of calculated annual change; (ii) comparing with data on the interannual variability; (iii) simulating the annual change of palaeoclimate in the last glacial period (18 thousand years ago).

The most important impacts of sea on climate revealed through numerical modelling are as follows: (i) maximum climate warming with an increasing CO_2 concentration in the wintertime Arctic due to the heat input from the ocean through the thinner sea ice (as a result of warming sea ice); (ii) the effect of albedo of the more extended sea ice 18 thousand years ago on the SAT, comparable with the impact of continental glaciers; (iii) possible reversal of the conventional relationship between the amplitude of the annual change of temperature and depth of the oceanic mixed layer with the sea ice dynamics taken into account, since a thinner mixed layer favours the strengthening of the wintertime sea ice, which causes a delay of data of numerical modelling confirmed, on the whole, these conclusions (Global..., 1992).

A great contribution to understanding the problem under discussion has been by the international programme MIZEX (MIZEX..., 1989), aimed at studies of the processes of mesoscale interaction among the atmosphere, ocean and ice cover in the marginal ice zones near the sea ice edge (MIZ), since the MIZ play a key role as factors determining the processes in the ocean of midlatitudes and the climate. The programme foresaw a study of such processes in two regions: in the Greenland Sea (Fram Strait) in deep waters, and on the continental shelf of the Bering Sea, with emphasis on observations in the Fram Strait, since (MIZEX..., 1989): (i) this region contributes most to water- and heat-exchange as well as other substances between the North Atlantic and Arctic Oceans; (ii) in this region various observational programmes have been already realized.

Analysis of the data available has shown that the processes of interaction of the key importance prevail in the periods from mid-July to late August and from mid-February to late March. Therefore two observational experiments were undertaken, in summer (1983) and in winter (1985), each lasting 6 weeks during the respective periods. The observational site (near the polar ice edge) covered a square of 200 by 200 km^2 with equal areas of ice cover and open water. This size of the site is enough to monitor the most substantial processes in the MIZ. So, for example, the longwave oceanic swell can penetrate the ice cover at a distance of 100 km causing the ice crash near the edge within a band of several tens of kilometres wide.

The thickness and size of the ice fields differ strongly from respective characteristics of the inner part of the ice cover, at least, at a distance of about 100 km from the edge. The diameters of the gyres "marked" by crashed ice are from 10 to 60 km. A considerable transformation of atmospheric boundary layer takes place within 30 km of the edge. The wind-driven motion of ice takes place at a distance of 100 km from the edge.

The choice of a six-week interval of observations ensures a representative coverage of the time scales of the processes in the atmosphere, ocean and ice cover. During this period several cyclones can pass across the test area, the gyres can form and disintegrate, oceanic fronts can deform, and the formation of upwellings and near-edge currents can take place.

The observational complex foresaw the use of a drifting station at the site, ships (icebreaker, research ship of an icebreaker type, conventional research ship), buoys, aircraft, satellites. Ships and aircraft played a decisive role in the accomplishment of the programme of observations of the atmosphere, ocean and ice cover (Table 1.2). In the

Table 1.2. Basic parameters of ice cover and ocean, remotely sensed with air-borne instruments (1, primary instruments; 2, auxiliary instruments)

Instruments / Parameter	Side-looking radar	Radio-altimeter	Micro-wave radiometer	Scattero-meter	Lidar profilo-meter	Wind sensor	Camera	IR scanner	IR profilo-meter	Thermo-sonde
Ice cover concentration	2	2	1	—	—	—	2	—	—	—
Type of ice	2	—	1	1	—	—	2	—	—	—
Wind over ice	—	—	—	—	—	1	—	—	—	—
Wind over water	—	1	—	1	—	1	—	—	—	—
Roughness										
SST	1	1	—	1	2	—	—	—	—	—
Temperature profile in the ocean	—	—	—	—	—	—	—	—	—	—

sub-programme of meteorological observations (over pack ice and open water) the emphasis was placed on the determination of wind stress over the ice and measurements of the surface heat balance components, as well as lidar profiling to retrieve the surface topography. The most important objects of the oceanographic sub-programme were gyres (their properties and dynamics), oceanic fronts, upwellings, tidal and inertial oscillations, small-scale processes (destruction of internal waves, convection), and roughness. In the ice cover observations the first priority is given to its dynamics and thermodynamics (inner stress, inertial effects, temperature field), small-scale processes near the edge.

The programme of observations in the Bering Sea has many similar features but also specific features determined by the following: (i) small depths (20–150 m); (ii) weak effect of sea currents in the MIZ (in contrast to the Greenland Sea); (iii) prevailing one-year ice (in contrast to the multi-year ice in the test area in the Fram Strait).

The wintertime increase of ice cover in December–Feburary with a maximum in March, and the springtime disintegration of ice in April–June characterize the dynamics of the ice cover.

One of the important and still unstudied features of the process in the MIZ is an appearance of divergence of the compact ice cover which leads to the deformation of the alternating bands of ice and water. In this connection, two major components of the summer (1985) and winter (1983) experiments were: (i) study of ice dynamics in the MIZ; (ii) observation of ice formation in the leeward polynyas. Analysis of satellite images of the ice cover has shown that the ice forms in the Bering Sea, as a rule, within several relatively small polynyas, where an extremely intensive heat exchange takes place between the water surface (releasing the heat) and the atmosphere, with subsequent formation of the ice cover.

Analysing the possibilities of studying the data on the sea ice dynamics as an indicator of early detection of anthropogenic climate change, Parkinson (1991) notes that, in general, such indicators should meet the following four general conditions: (i) a considerable variability ensuring a reliable record of the "signal"; (ii) possibility of regular global-scale observations; (iii) small and well-known natural variability (low level of "noise"); (iv) early changes with respect in other climate parameters (for example, the SAT).

From the viewpoint of the criteria mentioned, the sea ice dynamics cannot be considered an adequate indicator, since, for example, the ice cover is characterized by strong interannual variability (both on local hemispherical scales), and the observational series available are comparatively short. The phase relationships in the variability of sea ice and other climate parameters have been studied inadequately.

From the data of Zwally et al. (1976, 1983; International Conference, 1991, 1994) the amplitude of the regional interannual variability of the antarctic ice cover extent can reach 30%, and, on the average, for the Antarctic it is about 10%. The multi-year trend for the last 15 years has turned out, however, to be negligible. A substantial interannual variability of the ice cover extent has been also recorded in the Arctic. From the data for the period 1966–1986 the area of ice in the Barents Sea decreased by 25% in the end of the period, but no decrease of the maximum ice cover extent was recorded in late April for the same period (probably, it was a result of the ice cover thickness decreasing trend).

As has been mentioned earlier, the most impressive manifestation of anthropogenic impacts on the antarctic atmosphere is the formation of the springtime minimum of total ozone content ("ozone hole") due to the specific combination of circumpolar circulation and low temperature of the stratosphere, polar stratospheric clouds, and the input of chlorofluorocarbons (Kondratyev, 1989). Particularly interesting is the fact that the ozone hole event was predicted based on the seemed-to-be up-to-date stratospheric simulation models.

Also important are the anthropogenically induced changes in the arctic atmospheric chemistry, especially those connected with the arctic haze formation (Jaworowski, 1989, 1993, 1994; Jaworowski et al., 1990, 1992; Shaw, 1982a, b, c, 1985, 1988, 1993; Shaw et al., 1993). The destruction of ozone in the lower atmosphere caused by bromine and iodine compounds of, apparently, marine origin, has been quite unexpected; the contribution of anthropogenic factors cannot be excluded (Barrie, in International Conference, 1991).

Recent observations in the Arctic have revealed episodic decreases of ozone concentration in the Atmospheric Boundary Layer (ABL) from 30–40 ppb (by volume) to the level below the detection threshold during a period of less than one day in conditions of a very stable ABL stratification. The ozone depletion began at sunrise (with setting-on polar day) and continued during March–April. A high correlation was observed of the ozone concentration decrease with the level of concentration of gaseous bromine, which turned out to be much higher than during the remaining part of the year. To explain the respective cause-and-effect connection, it is necessary to answer two questions: (i) what is the source of filtered bromine (f-Br is total concentration of bromine, both gaseous and in particles, which can be measured with a filter sample); (ii) what are the ways of transforming gaseous HBr and Br-containing organic compounds (BCC) into more active forms of bromine.

MConnell et al. (1992) supposed that in the period of long polar night the concentration of Br (released by sea salts) in the snow cover grows with subsequent release of Br to the atmosphere in the form of Br_2 at sunset (the presence of sea salts on the snow surface can be explained by the permanent existence of polynyas and leads in the Arctic). In general, the gas-to-particle conversion of Br_2 into HBr or BOC should not cause a substantial ozone depletion in the ABL. One may think, however, that HBr and BOC are washed-out from the atmosphere by aerosol particles and ice crystals, and during this process the photostimulated heterogeneous reactions of "return" formation and release Br_2 to the atmosphere take place.

Such a cyclic process of the aerosol–atmosphere bromine exchange can maintain the level of concentration of Br atoms and BrO radicals sufficient to provide an agreement of the ozone depletion intensity with the observational data. The atoms of Br appearing with BrO photolysis can cause a fast (during 1–2 days) depletion of ozone in the process of subsequent catalytic reactions:

$$Br + O_3 \rightarrow BrO + O_2$$

$$BrO + BrO \rightarrow 2Br + O_2$$

The role of accelerators of the process of ozone depletion can be played by HCHO and CH_3CHO. To verify the hypothesis discussed, further laboratory experiments and field observations are needed (in particular, the vertical distribution of Br in snow cover).

As Schnell et al. (International Conference, 1991) show, a considerable photolytic destruction of ozone takes place around the summertime Antarctic in the atmospheric boundary layer. As for the arctic haze, during the last years its sources and ways of input to the Arctic as well as the chemical composition and size distribution of aerosol have become more clear, though the vital need for further studies remains.

Very interesting is the input of heavy metals to polar regions (Adamenko et al., 1991a, 1991b). Blanchet (International Conference, 1991) performed a numerical modelling of climate changes due to the arctic haze and showed that the effect of concentration of the soot component 0.2 μg m^{-3} (this exceeds the observed value by a factor of 5) was equivalent to the effect of a CO_2 doubling, the effect of the aerosol haze manifesting most in the regions of the Canadian arctic archipelago and the Greenland Sea.

The environment and the biosphere of the Arctic are characterized by the unique combination of such components as the polar ocean, ice and snow cover, glaciers, tundra, permafrost, boreal forests, and wetlands, each being a sensitive indicator of global changes subject to considerable variations under the slightest effects, manifesting themselves as variations in the solar radiation input, surface temperature, heat transport by the ocean, chemical composition of the atmosphere and the ocean, content and properties of atmospheric aerosol, etc. On the other hand, the processes in the Arctic can substantially affect climate changes. This refers, for example, to the role of permafrost and wetlands in the formation of biogeochemical cycles. Of particular interest are such problems as the role of methane as a greenhouse gas, and the biodynamics of the euphotic zones of the Arctic and Antarctic Oceans (it is, first of all, the dynamics of phytoplankton and, hence, the release of dimethylsulfide to the atmosphere). Apparently, the high latitude regions are parts of the globe where the cosmic factors (mainly, solar activity) affect significantly the environment and the biosphere.

An assessment of the role of the analysis of the interaction of the following factors: (i) nutrients fluxes; (ii) carbon accumulation; (iii) the dynamics of ecosystems and biological communities; (iv) hydrology; (v) permafrost; (vi) snow cover and glaciers; (vii) coastal processes; (viii) sea ice; (ix) oceanic circulation and formation of bottom waters; (x) atmospheric dynamics and heat balance; (xi) atmospheric composition; (xii) solar and geomagnetic effects. The necessity to take into account the interaction of the factors mentioned determines the priority of seven interdisciplinary problems:

(1)   The response of the arctic marine and continental ecosystems to external conditions. Interactive variations in the arctic ecosystems can manifest themselves, on the one hand, as the effect of changes in climate, world ocean level and other environmental parameters on the ecosystems and, on the other hand, as the effect of the dynamics of ecosystems (for example, through changes in the carbon cycle) on the environment and, probably, global climate.

(2)   The role of the arctic wetlands in the formation of biogeochemical cycles can manifest itself, first of all, in varying carbon accumulation which ought to have global consequences resulting from either assimilation or emission to the atmosphere of carbon dioxide and methane. In this connection studies of interaction between various cycles in the processes of the biosphere–atmosphere exchange in which the soil biology and permafrost play a special role are important. It is worth

noting that studies of connections between palaeoclimate and palaeoproductivity are particularly urgent.

(3)   The biotic dynamics at the ice cover–open water interface. The high bioproductivity near the ice cover edge in the regions of polynyas necessitates a study of processes determining the accumulation of plankton and dynamics of the nutrients input, which leads to the blooming of plankton.

(4)   The effects of the arctic glaciers and ice sheets on the ocean level in conditions of climate warming should be taken into account, from the viewpoint both of its contribution to the increase of the world ocean level and of the effect on the coastal zone and plankton blooming.

(5)   Special emphasis should be placed on the water cycle in the Arctic, having in mind the unique nature of the hydrological ecosystem controlled by a complicated complex of processes of melting and freezing of water. The soil moisture, the level of underground waters and river run-off greatly affect the nutrients production, gas exchange, bioproductivity, and other processes.

(6)   The chemistry and dynamics of the arctic atmosphere require serious attention in connection with global consequences which can cause changes in the concentration of greenhouse gases (GHGs) and stratospheric ozone, determined by a complex combination of chemical and photochemical reactions which depend on special features of the high latitude atmospheric general circulation (AGC).

(7)   The urgency of the problem of interaction in the ocean–ice cover–atmosphere system is determined by the key role of sea ice in the global-scale processes, especially from the viewpoint of the Earth's radiation budget (ERB) variability (Kondratyev et al., 1988a) and the formation of bottom waters and, hence, climate.

In connection with the problems mentioned, the Arctic (like the Antarctic) is important as a source of palaeoclimatic information contained in ice cores and bottom sediments, although interpretation of ice core data still contains serious uncertainties (Charles et al., 1994; Jaworowski et al., 1992; Jaworowski, 1994; Newell and Zhu, 1994).

Of particular interest are studies of the role of polar regions in the formation of the global carbon cycle. The arctic ecosystems are known to contain a quantity of carbon as organic substance of soil and, depending on climate changes in future, can be a powerful source of, or sink for, carbon. The contribution of the tundra and boreal forests to the global reservoir of carbon on land constitutes about 30%. The data of observations on the island of Tulik (Alaska) considered by Ochel (International Conference, 1991) revealed a tundra-to-atmosphere carbon flux reaching 3 g m^{-2} per day, which, calculated for the Alaska tundra, is on the whole equivalent to $(0.1–0.2) \times 10^9$ t of carbon per year and for the whole of Alaska 55 Gt of carbon per year.

Bearing in mind the estimates of the contribution to the methane global budget, Reeburg and Walen (International Conference, 1991) performed measurements of CH_4 flux at several locations of the Alaska tundra during 3.5 years and in the taiga (for less than one year). In the period considered (the fall of 1989) the taiga was always a sink for methane. The global estimates from these data gave the emissions of CH_4 by the tundra within the range 19–33 Gt yr^{-1}, whereas the contribution of the taiga as a sink for methane constituted 15 Gt yr^{-1}. The numerical modelling carried out by Bonan (International

Conference, 1991) has shown that boreal forests of the island part of Alaska should be a zone of carbon sink.

Kvenevolden (1988) analysed the role of vast reservoir of hydrates of different gases, including the GHGs (methane, nitrous oxide) "frozen" in permafrost (including the regions of shelf), which, in condition of climate warming, can intensively enter the atmosphere. The supplies of methane hydrate in permafrost have been estimated at 400 Gt of carbon. On the whole, the atmosphere can get up to 120 Gt of carbon per year at the expense of the Arctic. Since sea transgression has taken place in Eastern Siberia during the last 27 thousand years, it is possible that in this connection some amount of methane has entered the atmosphere. When discussing the problem of the role of polar regions in the formation of global changes, one should have in mind that two aspects of global ecology are of paramount importance (Kondratyev at al., 1988a; Kondratyev, 1990): (i) the anthropogenically induced redistribution of the heat balance components of the Earth as a planet (with emphasis on the problem of atmospheric greenhouse effect and its climatic impact); (ii) anthropogenically induced breaking of global biogeochemical cycles (primarily referring to carbon, nitrogen and sulphur).

Polar regions are a component of the global ecosystem. From the viewpoint of the Earth energetics, these are zones of the powerful sink of energy and their value in the formation of biogeochemical cycles is determined by the presence of high latitude ecosystems highly sensitive to the external forcings. The latter means, however, that the dynamics of the processes in polar regions can be the most reliable indicator of global ecological changes.

2

High latitude processes and global climate

2.1 SPECIFIC FEATURES OF HIGH LATITUDE CLIMATICALLY SIGNIFICANT PROCESSES

As a rule, the emphasis in research on numerical climate modelling has been laid on studies of climate in low and middle latitudes. The importance of processes in high latitudes has been underestimated. Two areas that have been a major focus are: (i) the response of climate to the growth of CO_2 concentration is supposed to be focused in the NH high latitudes; (ii) the deep-water circulation (convection) in the North Atlantic and in the Antarctic is the principal factor in the formation of thermohaline circulation. The presence (or absence) of such a circulation depends markedly on the salinity and heat balance of the atmosphere–ocean–ice cover system on the outskirts of the Arctic and Antarctic.

Of special importance are climatically significant processes in the Arctic such as: energy- and mass-exchange between the ocean and atmosphere, gigantic thermal inertia of the ocean, land–ocean temperature contrast, meridional heat transport from lower to higher latitudes in the upper (wind-driven circulation) and deeper (thermohaline circulation) layers of the ocean (Sandven, 1992).

An important aspect of the arctic climate dynamics are the results of numerical modelling demonstrating an enhancement in high latitudes of the northern hemisphere of the greenhouse-induced air surface temperature increase and climate warming in general (Manabe and Wetherald, 1980; Manabe et al., 1991, 1992). Kahl et al. (1993a, 1993b) have examined arctic temperature trends in four tropospheric layers during the period 1958–1986. The principal findings are as follows:

(1) The majority of the temperature trends are not statistically significant at the 90% confidence level. The results therefore indicate that a greenhouse-induced global warming "signal" in the Arctic is not detectable for the 1958–1986 period.

(2) Absolute station trends of 3°C per 30 years or greater were found, with both cooling and warming tendencies observed in all layers.

(3) There is considerable regional and seasonal variability in the sign and magnitude of the temperature trends. Winter trends in the 850–700 hPa layer, which according to

theory should be positive and large, are in fact negative over western Eurasia and eastern North America and Greenland.

(4)   Temperature trends at many stations in Eurasia and parts of Greenland are strongly sensitive to positive anomalies contained in the Massachusetts Institute of Technology (MIT) archive covering the period 1958–1963. Because of frequent errors contained in this archive, it was eliminated from the initial trend analysis. Many warming trends computed without these data become cooling trends when these early data are included. The present analysis is unable to conclusively determine whether the period 1958–1963 was in fact anomalously warm in Eurasia and Greenland. However, given the uncertain quality of the MIT archive, it was advisable to exclude it from the trend analysis at this time.

Thus, the observational results discussed by Kahl et al. (1993a, 1993b) show that an Arctic-wide "greenhouse warming" signal is not present during the 1958–1986 period. Because this conclusion is in a serious disagreement with numerical modelling results it means that present-day climate models are still not adequate.

Chapman and Walsh (1993) have undertaken an analysis of recent variations of sea ice extent (SIE) and surface air temperature (SAT) in high latitudes using both conventional and satellite observational data. It has been pointed out that the arctic sea ice variations for the past several decades (1961–1990) are compatible with the corresponding air temperatures, which show a distinct warming that is strongest over northern land areas during the winter and spring (average summertime SAT trend for the north polar region is practically absent). The temperature trends over the Siberian seas are smaller and even negative in the southern Greenland region, but SAT increase is maximal in subarctic latitudes of Alaska, northwestern Canada and the northern part of Eurasia. Statistically significant decreases of the summer extent of arctic ice are apparent in the sea -ice data, and new summer minima have been achieved three times in the past 15 years. There is no significant trend of ice extent in the Arctic during winter or in the Antarctic during all seasons.

Chapman and Walsh (1993) have emphasized that the strongest climate warming was observed in those land areas (northwestern part of the USA, northern part of Asia) and during those seasons (late winter–early spring) where and when the albedo feedback was the strongest. These results are important for solving the problem of "greenhouse signal" detection. It is very important in this context to accomplish detailed intercomparison between observations and numerical modelling for the atmosphere–ocean–ice cover coupled system.

On the basis of a regional model for studies of atmosphere–ice–ocean interaction in the Western Arctic Walsh et al. (1993) have shown that there is a resolution dependence in the land–atmosphere interaction that affects surface winds and precipitation. The ice formation rates simulated by a sea ice model are also dependent on the spatial resolution. While these resolution dependences might have been anticipated, it is perhaps surprising that it is detectable even in the comparison of simulations made at 21 km and 7 km resolutions. Resolution will have an even greater impact on simulations made with the coarser (50–100 km) mesoscale grid and especially on simulation made at GCM resolution (400 km).

The early results point to the importance of realistic surface–atmosphere coupling over the high latitude oceans. The absence of sea ice in the Bering/Chukchi region clearly had a major impact on the air temperatures simulated by the atmospheric model, and the temperature changes were sufficient to modify the field of sea level pressure, surface wind and precipitation even at 48 hours. Climatic impacts of ice reduction would clearly be substantial. Perhaps more importantly, the need for realistic ice–atmosphere coupling was indicated by the interdependence between the ice–ocean forcing of the atmospheric model, the simulated atmospheric winds, and the drift speed and open-water formation rates of the simulated sea ice. The immediate priority therefore is the implementation of a fully coupled atmosphere–ice–ocean model.

In connection with a special role of coupling between various components of the climatic system in the Arctic, Walsh and Crane (1992) have undertaken an intercomparison of climate modelling results for high latitudes with the use of five climate models which do not foresee an account of interaction between ocean and atmosphere. These models have been developed in the following institutions: Geophysical Fluid Dynamics Laboratory (GFDL), Goddard Institute for Space Studies (GISS), National Centre for Atmospheric Research (NCAR), Oregon State University (OSU), UK Meteorological Office (UKMO).

Surface air temperature and surface pressure (geostrophic wind) have been chosen as parameters which are most sensitive to interaction between the atmosphere and ocean (under Arctic conditions of special significance is parameterization of ice cover dynamics). In all cases results of control integration ($1 \times CO_2$) have been averaged over time periods which were not less than three last years of integration (the intercomparison is being complicated by a difference in the spatial and temporal resolution of the models). Table 2.1 data illustrate correlation between calculated and observed pressure fields.

Table 2.1. Correlation coefficient between calculated and observed
pressure fields, 70°–90°N

Month	UKMO	OSU	GISS	GFDL
DJF	0.848	0.387	0.414	0.909
MAM	0.792	0.398	0.532	0.908
JJD	0.378	0.497	0.356	0.568
SON	0.743	0.169	0.550	0.760

As can be seen from Table 2.1, there is a strong variability of the correlation coefficients. During non-summer months, of the most active ice cover dynamics GFDL and UKMO models happen to be the most reliable, while during summer months (especially in July and August) none of the models are adequate. Similar data for surface-air temperature are given in Table 2.2.

From the viewpoint of systematic errors the UKMO model looks the best (except summer months) which results from a reliable enough simulation of the surface heat balance. In case of the GISS model, almost all year-round differences (except spring

Table 2.2. Comparison between calculated and observed surface air temperature (°C), 70°–90°N

	Average absolute error (°C)				Average difference model-observation			
	UKMO	OSU	GISS	GFDL	UKMO	OSU	GISS	GFDL
DJF	4.26	8.42	4.28	10.10	−0.74	4.67	−3.03	−10.07
MAM	2.54	6.13	3.16	5.46	0.31	3.66	0.99	−5.46
JJA	3.15	2.09	2.91	2.58	−2.15	0.17	2.38	−2.39
SON	3.37	6.08	4.97	9.99	−0.22	4.36	−4.64	−9.95

months) of the order of 2–4°C take place, while the OSU model is characterized by the annual variations of the differences with the maximum absolute error of 8°C in November and December. The strongest deviations (up to −10°C) from observations exist in case of the GFDL model during the time period from October to February, which should result in the enhanced ice cover extent in winter.

It is a kind of paradox that the GFDL model which is the best in simulation of pressure field is the least adequate with regard to temperature simulation (a significant cause may be the lack of account being taken of leads and polynyas).

A significant disagreement between calculated temperatures, as well as between calculations and observations, is a result of inadequate simulation of thermodynamic interaction between the atmosphere and the ocean under the presence of ice cover. The further development of climate models requires the taking into account of: (1) more realistic ice cover dynamics, better simulation of ice cover transport, formation of leads and polynyas; (2) interactions in the atmosphere–ice cover system which are responsible for dynamics and thermodynamics of the ice cover; (3) parameterization of extended stratus cloudiness and its interaction with radiation.

Of increasing importance now is the numerical modelling of the interactive system "atmosphere–ocean–ice cover".

All hemispherical and global climate models take into account the processes in high latitudes, but relevant parameterization schemes are diverse. Numerical climate modelling of polar regions is seriously complicated by a number of factors. One of them is a radical difference between the climates of the Arctic (mainly, the ocean covered with ice, whose dynamics is governed by temperature and convective mixing) and the Antarctic (continental ice sheet up to 4 km high), with scarce observational data. Another factor is an inadequacy of grid models near the poles (the existence of singularity) leading to strong distortions of results (here spectral models are more useful).

Another difficulty is a reliable simulation of specific cloud conditions in high latitudes. With snow cover, the formation of extended cloudiness leads to a surface heating (independently of the height of clouds), whereas in the case of the snow-free surface the low-level extended cloudiness causes a cooling, and the upper-level clouds lead to a warming (the greenhouse effect of cirrus clouds).

The effect of the dynamics of permafrost on the hydrological regime is very important. Estimates show, for example, that with permafrost destroyed in the regions of Lena and Enissey rivers the run-off for these rivers will be halved, which will tell on the salinity regime of the arctic seas and the annual change of the extent of the arctic ice cover (the process of ice formation will slow down). Changes in the snow-ice cover albedo caused by contamination can play a substantial role. This effect must be particularly considered in analysis of the effect of a possible nuclear war on climate. More accurate consideration of the processes in polar regions is a fundamental aspect of further improvements on climate models.

To consider the problem of water exchange between the continental ice cover and the world ocean during the last century and the possibilities of forecasting climate changes and the ocean level during the forthcoming hundred years, a report was written (Glaciers…, 1985), where an analysis was made of grounds for the previous opinions that a climate warming due to an increase of CO_2 and other greenhouse gases will cause a continental ice melting and thermal expansion of the upper oceanic layer, which will lead to a considerable (possible, catastrophic) rise of the world ocean level. The report contains a discussion of the problem of water exchange between continental ice and the world ocean for the last century, as well as of possibilities of forecasting climate changes and sea level variations during the century to come. The general conclusion is that the ocean level is rising but the speed of rising is known only to an accuracy of coefficient 2 (from the beginning of the century it was 1–3 mm yr^{-1}), which is explained by both fragmentary observations (especially in the southern hemisphere) and difficulties in estimating the effect of tectonic and isostatic factors. Preliminary estimates of the continuous adjustment of the Earth to shrinking ice sheets (the Scandinavian and Laurentide ice sheet) gave values 1 mm yr^{-1} for the eastern coastline of the USA.

The SST has increased from the beginning of the century, probably, by $(0.6 \pm 0.3)°C$, but this quantity can be distorted by systematic errors. A certain contribution to the rising of the ocean level is made by melting mountain glaciers and small ice caps, but, apparently the share of the Greenland ice sheet in this contribution is negligible. As for the antarctic ice sheet, most likely, it is growing, taking water from the ocean. However, the rate of the oceanic mass variations is practically zero.

It is will unknown, whether the observed rise of the world ocean level may be explained only by thermal expansion of its upper layer. Calculations of thermal expansion are unreliable because of the neglect of the deepwater processes. The limited observations of temperature, salinity and density of the oceanic water during the last several decades did not reveal any statistically substantial variations in individual regions (continuous observations are needed to draw reliable conclusions). North of 50° N, a slight freshening of the Atlantic water masses took place after 1972, but it could be temporal.

In assessing the balance between water masses and the ocean level for present conditions, a decrease in the thickness of the peripheral parts of the Greenland ice sheet was discovered, as well as a thickening of its central part (Table 2.3). The uncertainty in the estimates for the Antarctic is explained by unreliable data on iceberg ejections, ice accumulation in some regions and ice melting beneath the shelf areas of the glaciers.

Physical processes governing the effect of climate changes on the Antarctic ice sheet are still inadequately studied. It follows from prognostic estimates that, apart from

Table 2.3. Estimates of the balance between the water mass of the ocean and its level

Ice mass	Equivalent change of mass (mm yr^{-1})	Contribution to the change of the ocean level (mm yr^{-1})
Glaciers and small ice caps	−1.2 ± 0.7	−0.5 ± 0.3
Greenland ice shield	0.02 ± 0.08	−0.1 ± 0.4
Antarctic ice shield	0.02 ± 0.08	−0.6 ± 0.6

rainfall intensification (with supposed climate warming), an additional contribution to growing ocean water masses can be made by melting small glaciers and the Greenland ice sheet, but the total rise of the ocean level will not exceed several tenths of a metre by the year 2100 (Table 2.4).

The estimates of the change in the ocean level connected with the Antarctic are rather uncertain. Apparently, it ought to be slight, but a new rise of the level by a few tens of centimetres is possible by 2100 at the expense of taking up the broken parts of glaciers.

Table 2.4. Estimates of the contribution by glaciers to changing the world ocean level at a climate warming caused by a doubled CO_2 concentration by the year 2100

Ice mass	Rate of changing the level (mm yr^{-1})	Total change of the level (m)
Glaciers and small ice caps	2–5	0.1–0.3
Greenland ice sheet	1–4	0.1–0.3
Antarctic ice sheet	−3–10	−0.1–1[a]

[a] The range 0–0.3 is most likely.

Further studies on the ocean level must be aimed at improving climate models, especially with regard to the role of sea ice and cloudiness, as well as processes in high latitudes; at a more reliable determination of the global change of the ocean level from observations; and at a more accurate estimation of the contribution by volume thermal expansion of the ocean. In conditions of the Antarctic, studies of heat transport in the ocean across the continental shelf round the Antarctic (especially in the regions of the Seas of Ross, Amundsen, and Bellingshausen, as well as the Weddell Sea), and oceanic circulation beneath large masses of shelf ice, as well as the effect of a possible climate warming on these processes are most important.

Since the numerical modelling revealed an increased sensitivity of the climate in polar regions to anthropogenic impacts, the monitoring of polar climate changes is particularly interesting from the viewpoint of an early detection of global climate variations

(Kondratyev, 1992a). For this purpose, different indicators can be used. For example, the 2–4°C permafrost temperature increase observed in the region of North Alaska is assumed to have taken place during the last century. A climate-warming indicator may be a retreat of small glaciers, which could have contributed markedly (to 50%) to the 10–150 cm rising of the world ocean level, which had taken place during the preceding hundred years. On the other hand, in some studies a possibility has been mentioned of intensified snowfalls in climate-warming conditions and, hence, of the growth of some glaciers on Alaska.

Though the climate warming should have resulted in globally decreased snow cover and sea ice cover extent, an analysis of satellite data accumulated during the last 20 years has not revealed such a trend. The estimates obtained by glaciologists from conventional data show that the Greenland ice sheet has had a stable mass (and thus, had been in an equilibrium), and the mass of the Antarctic ice sheet has been increasing. In this connection, satellite radar and lidar sounding is important as a source of information about the balance of the ice sheet masses. An analysis of the visible and near-IR images obtained from NOAA and DMSP satellites permits a monitoring of the glacier dynamics. The lake ice dynamics (the dates of the onset and melting of ice cover), resulting from the analysis of images in the optical and thermal regions, can serve as a sensitive indicator of regional climate variations.

Interesting new results have been presented at the Conference "The Role of the Polar Oceans in Global Change" held in Fairbanks (USA) on 11–15 June 1990 (International Conference..., 1991; Kondratyev and Kotliakov, 1991).

Based on the use of a 1-D thermodynamic climate model, Curry and Ebert (1989, 1992; Role..., 1990) assessed a possible effect of the intensified pollution of the arctic atmosphere, which can lead to the growth of concentration of condensation nuclei, subsequent increase of concentration, decrease of the sizes of cloud droplets, and changes in the ice cover radiation budget, which will cause an increase of the ice cover thickness. On the other hand, the ice thickness is sensitive to the frequency of occurrence and optical thickness of ice clouds in the lower troposphere, appearing in the cold seasons.

Curry et al. (1993) have shown that, owing to changes in surface albedo and temperature-associated changing cloud properties, there is a strong nonlinearity between cloud properties and surface radiation fluxes and therefore:

— no simple relationship exists between top of the atmosphere radiative fluxes and surface radiative fluxes;
— clouds have a net warming effect on the surface in contrast to clouds at lower latitudes;
— for annual average cloud amount less than 0.4, the cloud radiative forcing is positive (warming) over the entire annual cycle.

New results of numerical climate modelling are extremely important for analysis of observational data on climate parameters aimed at revealing the anthropogenic climate "signal". Having performed an integration for the period 200 years for the atmosphere–ocean system with account of the deepwater circulation of the ocean, Manabe (Role..., 1990) demonstrated the practical absence in the Antarctic of not only the warming enhancement widely discussed but also any climate trend in general (so great is the effect of the gigantic thermal inertia of the ocean). Similar calculations by Washing (International Conference..., 1991) for a period of 30 years, which generally speaking, is insufficient to reach a stabilized regime, revealed a vast zone of climate cooling in the region of the North Atlantic. Analysing the data of observations in Fairbanks, Bowling (Role..., 1990) has shown that the low S/N ratio (warming/interannual variability) in high latitudes determines a marked contribution of various SAT observational errors bringing forth the "parasite" trends (such was the nature of the considerable warming recorded in Fairbanks in 1970).

Gordon (Role..., 1990) notes that the SH ocean considerably contributes to the formation of waters which largely determine the present abnormally cold ocean. This situation is mainly determined by the existence of the antarctic glaciation. Two regions determine the formation of water masses affecting differently the global circulation of the ocean: (i) the antarctic zone of the formation of cold dense waters propagation along the sea bottom, intruding deep into the northern hemisphere; (ii) the polar frontal zone (within the circumpolar current) where the low-salinity water masses appear moving northward near the low boundary and within the bottom part of the thermocline. The extent of the regions of water mass formation depends strongly on the presence of polynyas and an intensive atmosphere–ocean interaction taking place in them.

Budd and Simmonds (International Conference..., 1991) performed a numerical modelling of the greenhouse climate changes, prescribing (in July) the 2°C SST increase in the tropics and the 6°C increase in high latitudes, as well as decrease of the antarctic ice cover extent by 7° lat., and assuming different values of the relative area of leads: 1% (control), 5%, 20%, 50%, 80%, and 100%. Particularly marked were changes in water cycle in the NH high latitudes. Naturally, in the regions of the low ice cover concentration the fluxes of latent and sensible heat strongly increased. An intensified evaporation brought forth increasing precipitation in the coastal Antarctic and "assimilation" precipitation in the polar cap, which led to a decrease of sea surface level. With the rain rate assumed to increase by 30%, it should cause the 2 mm yr^{-1} decrease of the level.

Having compared the SAT observation data from 30 antarctic stations for 30 years with satellite data on the ice cover extent (ICE), Walsh and Zwally (International Conference..., 1991) showed that, as a rule, large SAT anomalies were followed by ICE anomalies of the same sign, but this correlation was not the same in the different regions, in particular, from the viewpoint of the phase shifts (the ICE changes lag behind SST anomalies but not in every case). In the Antarctic a complicated pattern of SAT variations of different signs were observed, with an average trend equal to zero.

There is no doubt that the ice cover dynamics in the Antarctic (the area of seasonal ice is practically equivalent to the territory of the Antarctic) substantially affects the climate. However, the problem is that the SH sea ice properties are still poorly known (ice cover thickness, texture characteristics, etc.).

As Lange (International Conference..., 1991) notes, in contrast to the arctic ice, the one-year ice of the Antarctic is 0.97 ± 0.73 m thick, and the thickness of the two-year ice constitutes 1.23 ± 0.68 m. An analysis of observations in the Greenland Sea performed by Mysak and others (International Conference..., 1991) led to the conclusion that severe ice conditions took place in 1902–1920 and in the late 1940s, whereas the 1920s and 1930s were characterized by light ice conditions, as a rule.

Breitenberger and Wendler (International Conference..., 1991) discussed the effect of the atmosphere–ice cover interaction in the Antarctic on the formation of AGC, emphasizing the strong variability of intensity, sign and phase of this interaction, which manifests itself most in the Ross and Weddell Seas.

Particularly substantial is the variability of the snow cover extent: the difference between maximum and minimum values reaches 80%. Rapid changes in the snow and ice cover, connected with thawing, are one of the key factors of climate formation, manifested, mainly as the albedo feedback. Though the available information about the characteristics of permafrost is fragmentary, it can be assumed that this crysopheric component is the most sensitive indicator of climate.

In continuation of the analysis of the dynamics of snow cover thawing made by Foster (beginning from late 1960, the springtime thawing of snow cover in the tundra of North America started earlier than before, but the reasons for this trend are still not clear); Kodama et al. (International Conference..., 1991) note that the intensification of cloudiness in the process of global warming detains the snow melting, while the surface albedo decrease and the increase of atmospheric thermal emission produce an opposite effect.

As has been mentioned earlier, of great importance should be the effect of permafrost dynamics on the hydrological regime and the arctic biosphere in general. The change in the snow-ice cover albedo due to its pollution by depositing particles of the arctic haze can play a substantial role.

Let us now consider the contribution of various climatically significant processes in more detail.

2.2 SEA ICE, SNOW COVER AND CLIMATE

As Untersteiner (1982, 1984) notes, the most important large-scale problems connected with studies of the impact of sea ice on climate are: (i) study of the balance of ice masses in a given region determined by contributions of advection and ice formation; (ii) revealing the physical processes that control the ice cover extent (ICE) in conditions when it is not limited by the effect of land (the autumn-time and wintertime growth of the extent is largely determined by the ice formation and advection, whereas the effect of the summertime melting is a major factor in the sea ice retreat).

Because the annual change of ICE in the Arctic is characterized by a sharp maximum and "flat" minimum, it can be supposed that the prevailing winter conditions all the year round will not cause further increases in ICE (the extent is already quasi-balanced), but the setting of a hypothetical round-the-year summer regime of climate can lead to total melting of the arctic ice cover. The relationship between advection and ice formation which determines the location of the south boundary of the ice cover remains unclear. Still less studied and very complicated are small-scale and mesoscale processes in the zone of ice edge, where the boundary layers in both the atmosphere and the ocean are transformed.

The polar ice cover dynamics is of primary importance for the formation of climate (Kotliakov, 1968; Crane and Walsh, 1991): the arctic pack ice extent varies annually from 8 million km^2 in the summer to 15 million km^2 in the winter, and in the southern hemisphere the amplitude of the annual change of ice cover is much larger (within the range 2.5–20 million km^2). The spatial and temporal scales of the effect of the polar ice dynamics on climate are extremely large–from synoptic to decadal and larger, embracing the whole globe. Here both positive and negative climatic feedback occurs. The role of the polar cloud dynamics is also important; their amount varies from approximately 40% in the summer to 80–90% in the winter. From the viewpoint of the effect on the ERB in high latitudes, the outgoing longwave radiation plays the leading role, whereas in low latitudes and mid-latitudes the albedo feedback prevails.

The fundamental importance of the sea ice and climate problem has motivated a number of international discussions on the subject (Sea Ice and Climate..., 1988, 1990, 1992). As a result several field programmes have been initiated (Sea Ice and Climate..., 1992): international ice thickness monitoring programmes in the Arctic (AITMP) and Antarctic (AnITMP) with the use of submerged sonars and sea buoys in the Arctic (IABP) and Antarctic (IAnBP). Important contributions are being made by the Arctic Climate System Study (ACSYS) Programme and the Nansen Arctic Programme mentioned earlier, as well as by the effort to compile the global databank on sea ice properties (GDSIDB). A special place belongs to the Arctic Ocean Grand Challenge as a component of the European Committee for Ocean and Polar Sciences (ECOPS) Programme (Identifying..., 1994; The Arctic Ocean..., 1994).

Many programmes have been developed for environmental monitoring in Antarctica and the detection of environmental change (Weller, 1992b): International Transantarctic Expedition (ITASE); European Programme of Antarctic Ice Cover Analysis (EPICVA); First Regional Observational Experiment to Study the Troposphere (FRoST); Long-Term Ecological Research (LTER); Research of Antarctic Coastal Areas Dynamics (RACER); Research of the Ross Shelf Ecosystem (RISE); Project of Long-Term Monitoring of Sea Ice Cover Concentration (PELICOM); West-Antarctic Ice Sheet Studies (WAIS) and others. An important role is being played by such global programmes as Earth Observing Systems (EOS); Global Energy and Water Cycles Experiment (GEWEX); International Satellites Cloud Climatology Project (ISCCP); International Global Atmospheric Chemistry Programme (IAGAC); Joint Global Ocean Flux Study (JGOFS); World Ocean Circulation Experiment (WOCE).

A special place belongs to the ACSYS programme with the five main areas of activity (Report..., 1994):

— an Arctic Ocean circulation programme,
— an Arctic sea ice programme,
— an Arctic atmosphere programme,
— study of the hydrology of the Arctic region,
— modelling of basin-wide ocean–processes and ocean–ice–atmosphere interaction.

As far as the Antarctic is concerned, in the first, particular objectives of relevance from the perspective of physical climate research are to study (Report..., 1994):

— atmosphere–ice–ocean interaction,
— clouds and cloud feedback in the Antarctic,

— deep and bottom water formation,
— sea ice/ice shelf/ice sheet interactions.

The interaction of the ice and cloud covers in the Arctic is especially important (Kondratyev, 1986, 1988, 1991a, 1992a, b, c; Kondratyev and Zhualev, 1981; Kondratyev and Johannessen, 1993; Lappo et al., 1990; Marchuk et al., 1986). It means, for example, that the climate-warming-induced decrease of the polar ice cover will intensify evaporation and increase cloudiness. This will compensate for the albedo decrease taking place as open water appears. It is clear from such an analysis that the nature of the mechanisms controlling albedo feedback are very sophisticated.

Calculations based on a 1-D thermodynamical climate model have shown that in the regions where the interannual variability of the ice cover extent manifests itself most, the predicted ice cover thickness depends strongly on cloud conditions. This means that cloud variations may cause substantial variations in the ice cover thickness and extent. Reliable quantitative estimations of the role of these and other processes need more adequate meteorological and oceanographic information for polar regions than is now available.

It has already been shown that, studies of sea ice (during its maximum development it covers almost 10% of the world ocean's surface) must be of central concern. The climatic impact of substantial seasonal and interannual variations in the sea ice extent is determined by both its high albedo and heat moisture exchange between the atmosphere and ocean (the latter is particular strong in the presence of polynyas). Recent studies have led to the conclusion that the heat transport across the ice cover in the Arctic is more intensive than it has been considered earlier. No doubt, an adequate climate theory must be based on an interactive consideration of the processes of formation and dynamics of the ice cover. From the viewpoint of the numerical modelling of climate, analysis of the response of ice to climate changes and the effect of the ice on climate is very important.

The cryosphere is an important component of the climate system (Kondratyev, 1992a). This is testified to by the data in Table 2.5, which reflect the spatial and temporal variability of the cryospheric components.

Table 2.5. The spatial and temporal scales of the cryospheric
components

Ice sheets	16	10^3–10^5
Permafrost	25	10–10^3
Mid-latitude glaciers	0.35	10^4
Sea ice	23	10^{-1}–10
Snow cover	19	10^{-2}–10^{-1}

As is seen, the sea ice, land snow and permafrost play the leading role among the cryospheric components. A typical feature of the cryosphere is a fast variability of its components in the annual course. So, the extent of sea ice varies from 20×10^6 km in September to 2.5×10^5 km in the southern hemisphere and within the range $(15–8.4) \times 10^6$ km^2 in the northern hemisphere.

The variability of snow cover is more substantial: the difference between maximum and minimum values reaches 80%. Rapid changes in the snow and ice covers due to melting are one of the key factors of climate formation, whose major manifestation is the albedo feedback and the thermal insulation of the ocean (Cohen, 1994). Gleick and Rango (1994), and Rango and Martinec (1994) have assess hydrological implications of climate change, including snow cover dynamics.

Although the available information on permafrost is fragmentary, one may believe that this cryogenic component is the most sensitive climate indicator. Cox et al. (1993) have pointed out that, because the absence or presence of permafrost changes the snowmelt during spring and the development of vegetation in summer, it is very important to adequately parameterize permafrost dynamics in the climate models. Processes in frozen soils also influence significantly the hydrological cycle. It is very significant that the soil layer above the permafrost is the source of methane for the atmosphere. Cox et al. (1993) have developed a parameterization scheme to simulate permafrost processes which may be used in climate models. The scheme has made it possible, in particular, to correctly simulate methane emissions.

Walsh et al. (1993) have developed a fully coupled regional ice model of the atmosphere–ice–ocean system; atmospheric and sea ice models have been adapted to a western Arctic domain centred on the Bering Strait. It is clear that the key means of solving the problems mentioned is remote sensing from satellites. As experience has shown, in conditions of high latitudes, passive and active sounding in the microwave region is particularly important. Microwave passive sounding has made it possible, for example, to detect, within pack ice, vast regions of either open water or thin ice (the polynya in the Weddell Sea reaches a size of 300 by 1000 km^2, and it remains during the whole polar night).

Starting from November 1966, the NOAA (USA) has been issuing weekly maps of snow and ice cover distributions in the northern hemisphere, on a scale of 1:5,000,000,000 (in polar stereographic projection). Dewey (1987) processed such data for the period November 1966–October 1984 to draw monthly mean and annual mean climatologic maps of the cover frequency of occurrence (apart from the time period with snow cover) over the 89 by 89 grid points for the whole northern hemisphere, the tropics excluded. A grid cell was considered to be totally snow covered, provided 50% or more of the cell was covered with snow.

Table 2.6 characterizes the snow cover extent in different months and for a year, on the average. Naturally, the climatological maps drawn from satellite data, provide a more detailed presentation of the variability observed (especially in high latitudes and in mountains) than the maps based on the data of surface observation. According to satellite data for the last two decades, the snow cover onset in Asia and North America has taken place several weeks later than would follow from the data of more recent conventional observation.

Analysis of space-derived maps revealed a persistent northward shifting of the southern snow cover edge, but a simultaneous southward shift of the isoline of a 100% snow cover frequency of occurrence. This reflects a narrowing of the transition latitudinal zone, within which the frequency varies from zero to 100%, by 25–50%.

Table 2.6. Monthly mean snow cover extent (in 10^6 km^2) in the northern hemisphere during the period November 1966–June 1985

Extent month	Maximum (year)	Minimum (year)	Average	MSD
January	51.0 (1979)	41.2 (1981)	46.4	2.6
February	52.8 (1978)	40.1 (1970)	46.2	3.0
March	43.5 (1971, 1978, 1979)	37.0 (1970)	41.0	1.9
April	36.3 (1979)	27.6 (1968)	31.7	2.3
May	24.6 (1978)	17.2 (1968)	21.2	2.0
June	15.5 (1976)	7.5 (1984)	11.8	1.9
July	8.5 (1967)	3.6 (1982)	5.5	1.3
August	5.3 (1980)	2.7 (1970, 1984)	3.9	0.8
September	9.9 (1972)	2.6 (1970)	5.2	1.6
October	31.6 (1976)	7.4 (1968)	17.7	5.7
November	37.6 (1976)	29.5 (1979)	33.5	2.0
December	45.6 (1973)	37.1 (1969)	41.9	2.5
Year	28.3 (1978)	23.5	25.5	

An important part of the information relevant to polar ice sheets dynamics is precipitation data. Bromwich and Robasky (1993) have analysed precipitation trends over the ice sheets over the Antarctic and Greenland for the recent three decades. These data demonstrate that the precipitation rate over Antarctica appears to have increased by about 5% over a time period spanning the accumulation means for the 1955–65 to 1965–75 periods, while over Greenland it has decreased by about 15% since 1963 with a secondary increase over the southern part of the ice sheet starting in 1977. At the end of the 10-year overlapping period, the global sea-level impact of the precipitation changes over Antarctica dominates that for Greenland and yields a net ice-sheet precipitation contribution of roughly –0.2 mm yr^{-1}.

2.3 EXTENDED CLOUDINESS AND RADIATION

A very important climatically significant process in the Arctic is interaction between cloudiness and radiation (Kondratyev, 1992a; Marchuk et al. (1986)). By the initiative of the World Meteorological Organization, the Global Meteorological Experiment was carried out in 1978/79, aimed at obtaining a global database for one year, as well as focused field observational programmes during two Special Observing Periods (SOPs), from 15 February till 15 March and from 15 May till 15 June 1979 (First..., 1981).

Within the Russian national programme of the Global Meteorological Experiment (GME), during the SOP-2, aircraft measurements over the station North Pole (NP-22, the East Arctic) were made, as well as a sub-satellite experiment, flights in the regions of volcanoes of Kamchatka and Kuril Islands.

The choice of the region of the East Arctic has been determined by the fact that it belongs to the very important weather- and climate-forming regions of the globe. The presence in this region of the drifting station NP-22, operating within the programme POLEX-North, has made it possible to carry out in this part of the Arctic the complex observations using a flying laboratory. Studies near the Kuril Islands were the first attempt to investigate the radiation regime of the atmosphere in the region of active volcanism.

The aquatory of the eastern Arctic in June is characterized by the oceanic polar climate due to a special radiative regime of the atmosphere, by the presence of snow and ice covers, open water surface, specific cloud fields, and the effect of air masses from the continent and the Pacific Ocean. The extended cloudiness prevailing in late spring–early summer, was the principal object of studies.

2.3.1 The climatological and synoptic characteristics of the region of expedition

The Arctic basin gets heat from the southern latitudes, being a region of a powerful heat sink as a result of the negative radiation budget of the surface–atmosphere system. Therefore the Arctic basin affects strongly the atmospheric circulation over the adjacent continent and oceans.

In the warm seasons the Arctic is influenced by a weakly expressed low-pressure area, whereas weak anticyclones are observed over the adjacent regions of the North Pole Ocean, the Eastern Arctic, in particular, embracing also the sector of warm surface waters of the Pacific Ocean. Most often, the cyclones move to the Eastern Arctic from the Bering Strait. The map of the mean surface pressure for July shows a decrease of the average wind speed, the prevailing air flows with the clear-cut eastern component over the Eurasian coastline.

The monthly mean pressure in the region NP-22–Wrangel Island is about 1014 ± 2 hPa, the surface wind speed is small, averaging about 3–4 m s^{-1}. With the inversions typical of this region, the surface layer is quite isolated from the air masses, rapidly moving over it. This fact, together with the lacking topographic effects, determines a lower repeatability of strong winds near the surface. The air and surface temperatures in the region NP-22–Wrangel Island in June were 0–1°C and −1°C, respectively.

The special features of the formation of heat balance are of primary important in understanding the weather and climate variability. The surface energy balance constituents in June in the Arctic are characterized by the prevailing effect of radiative component. So, for example, the shortwave radiation, absorbed by the surface, constitutes 13 W m^{-2}, the downward and upward longwave radiation is 28 and 32 W m^{-2}, respectively. The latent heat fluxes due to the phase transformations of water and turbulent heat exchange with the surface; the heat exchange with deeper layers of the ocean are 10, 11, and 65 W m^{-2}, respectively.

Thus, diabatic processes are very important for the formation of the climatic conditions of the Arctic. The diabatic impact favours cyclogenesis in air masses moving over warm surfaces and anticyclogenesis over cold surfaces.

The summertime melting of pack ice leads to the formation of stable fog and low-level cloudiness observed for 100 days and longer. Usually, they are caused by an advection of relatively warm and moist air masses over the melting ice or cold water. The fog is not observed when the wind speed exceeds 10 m s^{-1}.

The repeatability of extended stratus clouds and fogs in the summertime Arctic varies from 50% to 80% and, on the average, they cover up to 70% of the area of the ocean outside the polar circle, that is, the lower cloudiness and fogs are phenomena of not only regional but also planetary scale significance.

Two types can be mentioned among the variety of fogs: advective and radiative. The radiative fogs are formed, as a rule, due to the radiative cooling of the surface and the adjacent air layers in the presence of either moist air and (or) aerosols in the sub-inversion layer, 0–300 m. Most often, radiative fogs are formed at a wind speed of about 1–3 m s^{-1} and temperatures from –20 to +20°C. They are characterized by the existence of the zone of wind intensification (mesojet) near the top of the fog, by typical vertical profiles of the turbulent mixing coefficients and water content.

The advective fog appears as a result of warm air masses moving onto a cooler surface (water–ice, land–sea).

The formation of stratified clouds in the boundary layer takes place in three stages: (i) water evaporation; (ii) the upper transport of water vapour and its accumulation beneath the inversion; (iii) cooling and condensation. Near the upper boundary a temperature inversion begins and a mesojet is observed; the average radius of droplets and water content increase with height. Here a powerful longwave cooling and increasing radiative heating due to shortwave radiation are observed, too.

The weather in the test area during the whole period of observations (10 days) was determined by the impact (from the north) of the high-pressure area in the arctic water basin, and cyclones moved along the eastern coastline of Russia. The synoptic situation considered was typical of this season in the Arctic.

2.3.2 The aim and the logistics of the field experiment

As a rule, studies of cloud–radiation interactions in the Arctic were carried out in the following four directions: cloud climatology; the radiative regime of the surface or the Earth-atmosphere system from the ground-based and satellite data, respectively; aircraft studies of micro and macro parameters as well as radiative properties of clouds; assessments of the heat balance components for the ocean–atmosphere system based on aerological observations. Studies of the multiparametric interaction of the radiative heat flux divergence, water phase transformations, and sensible and latent heat transport require complex surface, aircraft and satellite observations.

The major goals of the field study were: simultaneous measurements from the aircraft IL-18, the satellite Meteor-2-4, and the drifting station NP-22; sounding of the atmosphere over the ice and water in cloudless and cloudy weather; prolonged flights during 24 hours, bearing in mind an analysis of the process of evolution of clouds using the data of aerological sounding of the atmosphere at the stations NP-22, Cape Schmidt, Wrangel Island, Peveck, and Uellen.

The airborne instruments used in the experiment are given in Table 2.7, and, as can been seen, are not completely adequate, to meet the requirements of the information needed to study the processes of the cloud–radiation interaction. During the SOP-2 the station NP-22 drifted north of Wrangel Island near the location with coordinates 45° N, 177° W.

Table 2.7. Instruments carried by the MGO turbojet IL-18 flying laboratory

	Instrument	Parameter to be measured	Range of measurements	Error
1.	Thermohygrometer	Temperature	40–50°C	5%
		Pressure	760–300 hPa	10%
		Humidity	5–100%	5%
		Wind speed		5%
2.	Water gauge	Water content	0.7	20%
3.	Aerosol complex	Size distribution and chemistry of particles	0.2–20 μm	30%
4.	Lidar	Backscattering coefficient	0.63; 1.06 μm	10%
5.	Actinometric complex	Shortwave and longwave radiation fluxes	0.35–3.0 μm	6%
6.	Spectrometer	Shortwave radiation fluxes	0.35–0.95 μm	8%
7.	Indicatometer-albedometer	Angular distribution of brightness	0.5–1.9 μm (11 filters)	6%
8.	Radiation thermometer	Radiative temperature	10–12 μm	
9.	Camera	Images	0.3–0.8 μm	5%
10.	Navigation instrumentation	Navigation, parameters		

The aircraft sounding in a cloud-free atmosphere over the ice and water was made in near-noon hours with the "platforms" at the heights 200, 500, 1350, 2850, 4200, 5500, 7200, and 8400 m. In the presence of clouds, additional "platforms" were used over the cloud top (above 300 m with respect to the top), inside the cloud and under it. The atmospheric sounding at Kamchatka and near the Kuril Islands was made at the same levels on the lee side and windward side with the fumarole activity of volcanoes over the homogeneous surface of water and clouds, of volcanoes Alaid and Chepurachek.

During the sub-satellite flights the atmospheric sounding was made so that at a sub-satellite point over the NP-22 the aircraft made a "platform" at a height of 8.4 km, which enabled one to obtain simultaneous data about the radiative properties of the atmosphere at three levels and to assess the radiative regime of the Earth–atmosphere system based on synchronous aircraft and satellite data.

2.3.3 Results of aircraft measurements of the characteristics of radiation fields and cloudiness

During the expedition 15 flights were made: on 21–28 May at Kamchatka and in the region of Kuril Islands, and from 1 to 7 June 1979 in the eastern sector of the Arctic in cloudy and cloudless condition over the water, ice and NP-22.

Based on radiation measurements of the shortwave radiation fluxes, the radiative characteristics of the surface–atmosphere system, albedo, A, over the weakly reflecting water surface of the sea of Okhotsk increases with altitude, H, from 0.2 km to 8.4 km with $\Delta A/\Delta H \approx 0.004$ km^{-1} for the solar height $h_\Theta = 44°$. Over the Chukchi Sea for $h_\Theta = 35°$ and a thin haze $\Delta A/\Delta H \approx 0.006$ km^{-1}. Earlier data from the CAENEX programme over the Azov Sea for $h_\Theta = 80°$ gave $\Delta A/\Delta H \approx 0.009$ km^{-1} and in the case of the dust outbreak from Sahara 0.037 km^{-1}. The albedo of the Earth–atmosphere system over the ice decreases with altitude in the altitude range 0.2–8.4 km: $\Delta A/\Delta H \approx 0.01$ km^{-1}, this decrease being most pronounced in the lower 200-m layer. For example, the albedo of ice from the data of NP-22 was 0.77–0.79, and at a height of 200 m it was 0.70–0.72, especially in the presence of the cloud layer located above. The albedo of the system in a cloudy atmosphere over the water surface decreases with a gradient of about 0.003 km^{-1}, and over the ice about 0.01 km^{-1}. In the presence of an aerosol layer or haze the vertical profile of the Earth–atmosphere system albedo can be more complicated. The ice albedo as a function of ice concentration and type varies from 0.42 to 0.74, from measurements at an altitude of 200 m.

An intercomparison of the cloud albedo values at an altitude of 1 km over the water and ice has shown that the increase of the cloud albedo due to the effect of the surface located below clouds can constitute 0.2 to 0.3. The major difference in the albedos of stratified clouds over the water and ice is connected with different albedos of the surface and, to a lesser extent, it is determined by the effect of the phase state of clouds and by the different relationship between the liquid and solid phases of cloud particles.

The phase state of clouds is more substantial for clouds of middle, and especially upper, levels. The presence of large crystals in the cloud leads to a stronger forward scattering, which determines a lower albedo of such one-layer clouds compared to lower clouds (Ac, As ~ 0.5–0.65, Cs ~ 0.15–1.30).

The albedo of altocumulus and altostratus, as well as cumulostratus clouds over the ice and underlying layer of stratus clouds is characterized by the values 0.60–0.76, 0.52. The albedo of fog over the water for the thickness of the fog layer $\Delta H = 200$ m is 0.39; the presence of the fumarole plume over it (i.e. the presence of hydroscopic compounds: water vapour, halogens, sulfur dioxide, and carbon dioxide, ammonia, hydrogen sulfide, chlorine compounds) leads to the increase of albedo by 0.03–0.04, but at an altitude of 4 km this difference is within the error of albedo measurements.

Clouds in the eastern part of the Arctic were characterized by values of water content 0.05–0.15 g m^{-3}, with average values of the attenuation coefficient γ (km^{-1}) differing from the respective values over the European Territory of Russia (Table 2.8).

Table 2.8. Average values of the size of cloud particles, $\bar{r}$, liquid water content, $\bar{w}$, attenuation coefficient, $\bar{\gamma}$, and inhomogeneity scale, $\bar{x}$

Type of clouds		Characteristics			
		$\bar{r}$ (μm)	$\bar{w}$ (g m^{-3})	$\bar{\gamma}$ (km^{-1})	$\bar{x}$ (m)
St	Arctic	4.4	0.15	41.7	1.57
	ETR	4.5	0.31	42.6	1.62
Sc	Arctic	4.0	0.1	24.9	0.49
	ETR	4.7	0.23	47.9	1.47
As	Arctic	—	0.07	4.4	—
	ETR	—	0.28	25.5	1.15
Ac	Arctic	4.7	0.20	20.5	0.82
	ETR	50–100	—	22.5	—
Cs	Arctic	80–150	—	—	—

Based on the technique of integral parameters, with the average water content and attenuation coefficient of stratified clouds known, the average diameter of cloud droplets and their number density was estimated at 3.9 μm and 30 cm^{-3}, respectively, which was below the respective data for the European Territory of Russia (ETR), that is, the cloud in the Arctic, especially the stratocumulus ones, are thinner and particles are smaller than over the ETR.

Knowing the geometric thickness and the average attenuation coefficient, one can determine the optical thickness of clouds, using the formula

$$\tau = \sum_{(i)} \gamma_i \Delta H_i$$

and draw the dependence of the albedo of stratified clouds over the ice on their optical thickness.

It follows from the results obtained that the albedo of stratus clouds over the ice at a lower Sun elevation is higher than over land, water, or town (the albedo of the surface of water, town, suburban zone and ice was, respectively, 0.04, 0.20, 0.13, and 0.70; these are the earlier CAENEX results).

The absorptance of stratified clouds is maximum over town, which is connected with the presence in the cloud of an optically active (absorbing) aerosol as well as with the changing size distribution of cloud particles over town under the influence of hygroscopic condensation nuclei. The absorptance of clouds over the water and ice are nearly the same (within the error of measurements) and constitute 2–7% for stratified clouds with ΔH not more than 500–600 m. The rate of the radiative heating of lower clouds varies from 0.1 to 0.8°C h^{-1}, that of longwave cooling from 0.1 to 0.5°C h^{-1}.

An analysis of the vertical profiles of radiative heat flux divergence in cloudless and cloudy conditions over the NP-22 for 4 June 1979 and 7 June 1979 showed that the atmosphere in the clear sky weather was characterized by a sufficiently high transparency 0.80 ± 0.02, aerosol optical thickness 0.082 at a wavelength of 0.4 μm, and water content 0.5–0.6 "cm". The aerosol number density at the surface level from the data of NP-22 was 0.4 cm^{-3}, at an altitude of 200 m was 0.2 cm^{-3}, and at altitudes above 4 km it was negligible small for the particle sizes from 0.2 to 20 μm. In the presence of clouds the total equivalent water content of an air column from the aerological data was 1.2 "cm", the relative humidity at the surface level was 96%, and within the cloud 70%, with average liquid water content 0.10 g m^{-3}. The vertical profile of the radiative heat flux divergence due to the shortwave (SW), longwave (LW), and total (T) radiation is charac- terized by a maximum in the lower 500-m layer and in the presence of clouds in the lower 200-m layer. A small maximum in the layer 2.8–4.2 km is connected with the effect of a humid haze at this altitude. The total radiative heat flux divergence decreases with altitude, and above the level 2.85 km the radiative cooling prevails.

The simultaneous measurements of brightness temperature T_B in the atmospheric window 10–12 μm and temperature of clouds, as well as longwave radiation fluxes at the top and bottom levels of clouds made it possible to estimate their emissivity using the formulas:

$$\varepsilon = \frac{B(T_B)}{B(T)}; \qquad \varepsilon = \frac{E_{e1} - E_{e2}}{E_{b1} - E_{b2}}$$

where $B(T_B)$ and $B(T)$ are blackbody emissions for the respective temperatures, E_{e1} and E_{e2} are equivalent longwave radiation fluxes at the boundaries of the cloud layer; E_{b1} and E_{b2} are blackbody emission fluxes at the respective levels.

Calculations based on the observational data for the arctic clouds have shown that the emissivity of lower clouds in most cases is below unity and varies from 0.65 to 0.85. For mid-level altocumulus and altostratus clouds it constitutes 0.35–0.50, and for cirrostratus clouds 0.10–0.25.

Lidar measurements performed with the airborne L-1M lidar enabled one to obtain the vertical profiles of the atmospheric backscattering coefficients at the wavelengths 0.53 and 1.06 μm. Table 2.9 contains the data for four altitudes, H, in the Arctic, over the Central Asian desert (Repetek), as well as in the presence of stratiform clouds of the lower (St) and upper (Cs) levels. It also gives the estimates of the molecular scattering coefficient ρ_m at the altitudes 0.5 and 0.3 km for the respective wavelengths. Naturally, the backscattering coefficient at high altitudes in the cloudless atmosphere is close to the molecular one. The backscattering coefficient constitutes approximately 40 km^{-1} for St and 1 km^{-1} for Cs. A change in the ratio of the backscattering coefficients measured at two wavelengths

$$B_\pi = \frac{\rho_\pi (\lambda = 0.53 \ \mu m)}{\rho_\pi (\lambda = 1.06 \ \mu m)}$$

Table 2.9. The vertical profiles of the backscattering coefficient, ρ. 10^{-3} km^{-1} sr^{-1} at wavelengths $\lambda_1 = 0.53$ μm and $\lambda_2 = 1.06$ μm.

H (km)	Repetek		Arctic		St		Cs	
	1	2	1	2	1	2	1	2
0.5	10	5	7	3	5×10^3	10×10^3	20	5
4.2	7	2	5	1	4	0.7	6	3
2.8	6	1	4	0.7	3	0.6	4	1
7.2	4	0.7	2	0.4	2	0.3	1.0×10^3	0.07×10^3

$\rho\lambda_{1,\mathrm{m}} = 1.6 \times 10^{-3}$ km^{-1} sr^{-1}

$H = 0.5$ km

$\rho\lambda_{2,\mathrm{m}} = 1 \times 10^{-4}$ km^{-1} sr^{-1}

$\rho\lambda_{1,\mathrm{m}} = 0.7 \times 10^{-3}$ km^{-1} sr^{-1}

$H = 8$ km

$\rho\lambda_{2,\mathrm{m}} = 0.45 \times 10^{-4}$ km^{-1} sr^{-1}

is determined by the size distribution of cloud and aerosol particles and varies between 1.0 and 16. This ratio grows with the increasing size of particles in the aerosol layer, liquid and crystal clouds.

Table 2.10 shows the data on the spectral albedo and coefficient of reflection anistropy of ice covered with snow, at the surface level and at altitudes 0.2 and 8.0 km.

Table 2.10. The spectral albedo (numerator) and the coefficient of reflection anisotropy of the snow-covered ice at different altitudes, at the Sun elevation $(26 \pm 4°)$

H (km)	λ (µm)					
	0.51	0.99	1.13	1.24	1.38	0.3–3.0
0	0.90	0.65	—	—	—	0.77
0.2	$\dfrac{0.82}{1.04}$	$\dfrac{0.70}{1.10}$	$\dfrac{0.68}{1.10}$	$\dfrac{0.52}{1.19}$	$\dfrac{0.40}{1.16}$	0.68
8.0	$\dfrac{0.74}{1.05}$	$\dfrac{0.63}{1.08}$	$\dfrac{0.51}{1.05}$	$\dfrac{0.42}{1.08}$	$\dfrac{0.18}{1.09}$	0.56

The ice-snow albedo in the visible is higher than the total albedo and markedly decreases with altitude and in the IR. The snow reflection anisotropy grows with wavelength and altitude. The reflection anisotropy for clouds and water surface as compared with the snow-covered ice grows markedly (approximately by factors of 2 and 5, respectively) at the same Sun elevation. Thus, the Arctic is characterized by surfaces which reflect radiation non-isotropically, and this fact should be taken into account, especially when determining the shortwave outgoing radiation fluxes and the radiation budget of the Earth–atmosphere system.

Note should be taken that in some cases, under certain respective synoptic situations in the Arctic, an increase of the aerosol content can be observed as well as a decrease of the atmospheric transparency coefficient down to 0.6 in the presence of large-scale forest fires in Siberia, eruption of volcanoes at Kamchatka, outbreaks of anthropogenic aerosol from the lower latitudes and dust-sand outbreaks from the Gobi desert.

Deposition of small (~ 0.1 µm), strongly absorbing particles with a mixing ratio of only 1 ppm^{-1}, can reduce the albedo of snow and ice in the visible by 5–15% as compared to the albedo of pure snow ($\sim 96\%$), especially in the visible (Table 2.10).

In the IR the ice (or snow) is a strong absorber by itself, which determines a sharp decrease of albedo in this spectral region. The values of the real and imaginary parts of the refraction index for ice at some wavelengths are given in Table 2.11, and they are much less than the respective values of the imaginary part of the refraction index for minerals ($n_2 = 10^{-2}$–10^{-4}), products of volcanic eruptions ($n_2 = 10^{-1}$–10^{-4}), and soot carbon ($n_2 = 0.5$–1). Therefore the presence of soot in concentrations of only 0.3 ppm is

Table 2.11. The real n_1 and imaginary n_2 parts of the ice-snow refraction
index at $t = -7°C$ from 0.3 to 28 µm

λ (µm)	n_1	n_2	λ (µm)	n_1	n_2
0.3	1.380	3.95×10^{-9}	1.3	1.291	1.25×10^{-5}
0.4	1.319	1.30×10^{-9}	1.6	1.263	3.17×10^{-4}
0.5	1.312	3.20×10^{-9}	1.9	1.275	4.25×10^{-4}
0.6	1.309	9.75×10^{-9}	2.2	1.265	3.88×10^{-4}
0.7	1.306	3.35×10^{-8}	2.5	1.225	8.50×10^{-4}
0.8	1.304	1.13×10^{-7}	2.8	1.084	2.80×10^{-2}
0.9	1.301	1.10×10^{-7}			
1.0	1.299	1.93×10^{-6}			

sufficient to explain the decrease of the ice and snow albedo in the visible as well as a
more rapid melting of ice in the Arctic.

Based on the accomplished observations in the Arctic, the following conclusions can
be drawn:

(1) The albedo of the surface–atmosphere system decreases over the weakly reflecting
     (water) surface and significantly increases over the strongly reflecting (snow, ice)
     surface and depends on the presence of aerosol haze.
(2) The albedo of the ice of different concentration and type can vary from 0.4 to 0.8 as
     observed from the altitude 0.2 km and is at a maximum in the visible region of the
     spectrum.
(3) The albedo of clouds in the Arctic is markedly higher than the albedo of clouds of
     the same optical thickness over city, land and water surface over the ETR and
     depends on the surface albedo as well as on a number of micro- and macro-
     parameters.
(4) The presence of hygroscopic particles as a result of fumarole activity of volcanoes
     in the cloudy atmosphere raises the albedo of clouds, but their effect on the radia-
     tive properties is small as compared to the impact of anthropogenic hydrophobic
     aerosols (chiefly, soot).
(5) The net radiative heat flux divergence is at a maximum in the lower 500-m layer,
     especially in the presence of lower clouds; it decreases with altitude and is charac-
     terized by cooling in the atmospheric layers located above clouds.
(6) The emissivity of clouds (even at a low altitude) in the Arctic is below unity. This
     is connected with the prevailing crystal phase in clouds and is confirmed by a
     considerably different backscattering coefficient as compared to the attenuation
     coefficient in clouds.
(7) The Arctic is characterized by a surface with a strong reflection anisotropy.
(8) The Arctic, together with the Antarctic, are of all regions the least subject to the
     effect of anthropogenic factors, but in the case of an outbreak of optically active
     aerosols, its effect can lead to a faster snow and ice melting and, hence, a more
     substantial warming.

2.3.4 Numerical modelling of the dynamics of extended cloudiness and its interaction with radiation

One of the problems of studying clouds in the Arctic was an analysis of the dynamics of the formation and development of fog and clouds over the ice during the polar day over the NP-22. The periodicity of soundings was 6 and 12 hours. The complex of surface and airborne instruments was not completely adequate to meet the requirements for studying the evolution of clouds. Therefore the results of numerical modelling of the development of clouds were compared, mainly, with the data on radiation, optical, micro- and macro-characteristics of clouds in the course of their temporal evolution with the purpose of assessing the contribution of radiation to the energetics of the cloud atmosphere (First..., 1981).

Different heat flux divergences in the cloud layer can be considered in general to be based on the heat balance equation. So, for example, the rate of change of the humid entropy of the cloud mass per unit time can be expressed through the rate of change in the pseudo-equivalent potential temperature Θ_e, which is

$$\frac{\partial \Theta_e}{\partial t} = -v\Delta\Theta_e - \overline{w}\frac{\partial \Theta_e}{\partial z} - \frac{\partial}{\partial z}\overline{(w'\Theta'_e)} - \frac{\Theta_e}{c_p T}\varepsilon_r$$

where v corresponds to the horizontal wind speed, w is the vertical wind component, ε_r is the radiative temperature variation and $\overline{w'\Theta'_e}$ is the vertical turbulent heat flux (Θ_e is pseudo-equivalent potential temperature). An advective term $v\Delta\Theta_e$ and the contribution of large-scale downward motion $\overline{w}(\partial\Theta_e/\partial z)$ (whose value can, apparently, explain the difference of clouds over the continent and the ocean) are very difficult to estimate precisely, both based on the observational data and using the modelling results. Therefore usually a 1-D time-dependent model is considered on the basis that it approximately corresponds to a 2-D stationary model of cloud dynamics.

Three components correspond to the sensible heat flux $\overline{w'\Theta'_e}$: at the expense of the change of potential temperature $\overline{w'\Theta'}$, water vapour $\overline{w'r'_v}$ and droplet water $\overline{w'r'_d}$, which, together with the radiative heat flux divergence, determine the energetics of clouds. Therefore, it is important, based on calculations and observational data, to assess the role of these components in the evolution of extended clouds in the Arctic and to compare, where possible, the result of calculations and observations.

In the numerical modelling of the process of formation and evolution of clouds in the atmospheric boundary layer (ABL) with air masses moving from the continent to the ocean covered with ice, the same numerical model as in the modelling of low clouds and fogs over the land was used (Kondratyev et al., 1992a; Marchuk et al., 1986).

The model includes an equation of the ABL dynamics, the transfer equations for heat, water vapour, longwave and solar radiation, the latter being calculated in a two-stream approximation. In this model the microphysical technique has been used to calculate the characteristics of cloudiness: oversaturation of water vapour and spectrum of droplets, as well as content, droplet concentration, average size of droplets, and visibility have been calculated.

Table 2.12 shows a block-diagram of the model. Each group of the processes in the ABL is simulated with an equation system of its own. In the numerical modelling a

Table 2.12. The model block-diagram

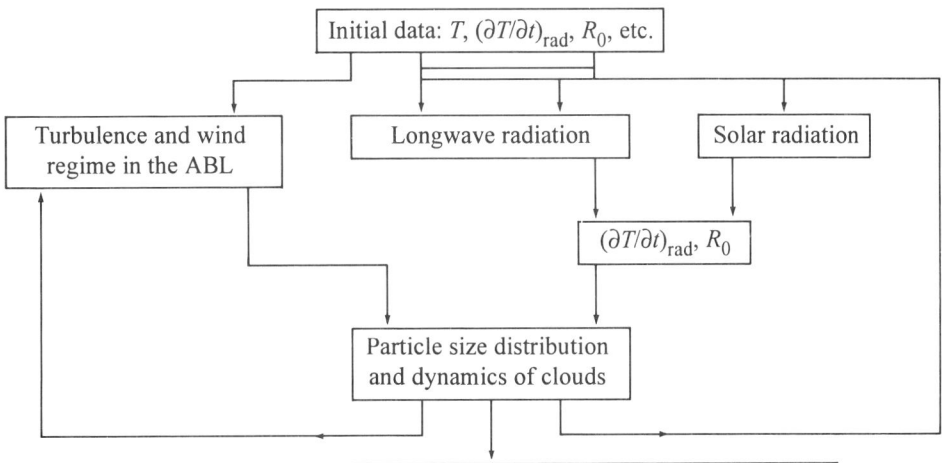

splitting technique has been used. The blocks in Table 2.12 correspond to the splitting into partial problems: arrows show the calculation succession, and the symbols beneath them show the exchange between them.

The synoptic situation during the experiment in the course of aircraft and aerological soundings was characterized by warm air masses covering a cooler ice surface, which led to the clear-sky weather change and formation of an advective fog, and its subsequent transformation into a stratus cloud.

An analysis of the vertical profiles of temperature, t, and relative humidity, U, in clear-sky weather, at the moment of the formation of fog, in 48 hours, and a stratus cloud in another 24 hours, showed that the process of cloud formation had been followed by a decrease of air temperature, raising of the upper boundary of inversion of temperature and relative humidity, which coincided with the cloud-layer top, with the growth of geometrical thickness of cloud from 200 to 300 m. The wind speed, according to the data of the airborne navigation instruments, was 3 m s^{-1}, the turbulent exchange took place when clouds appeared, the water content of fog averaged 0.06 g m^{-3} and that of cloud 0.11 g m^{-3}. With a radius of droplets of about 6 μm the content of precipitatable water in the surface layer in conditions of clear skies, fog and stratus clouds, was 0.29 "cm" and 0.61 "cm", respectively. The coefficient of turbulent mixing and vertical speed pulsations constitutes about $15 \text{ m}^2 \text{ s}^{-1}$ and 1 m s^{-1}, with a relative error of 50%. The surface temperature was $-2°C$, with the prevailing northeasterly winds.

Thus, a situation was modelled when a warmer air mass moved onto a cold surface, with the use of prescribed initial data. The melting ice at a temperature of $-2°C$ was taken as an underlying surface, initial relative humidity 80%, ice albedo 40%, geostrophic wind speed 10 m s^{-1}, and Sun elevations 20–35°, characteristic for the polar day. The temperature was considered to decrease with altitude with a lapse rate of $6° \text{ km}^{-1}$, with constant relative humidity.

The numerical modelling has shown that when the air masses move over ice, their cooling and sublimational sinking of water vapour onto the ice take place, and, hence, the temperature and specific humidity in the lower 300–440-m layer decrease in time, and their inversions form. However, the cooling rate exceeds that of water decrease, and 4 hours 40 minutes later a cloud forms in the layer 150–180 m. The mechanism for its creation is the same as for the formation of clouds over land covered with snow, then the downward water vapour transport follows. Its boundaries are 70 m and 360 m, its maximum water content (0.17 g kg^{-1}) is reached in the upper third of the cloud. The average radius of droplets is 5–6 m. This picture coincides with the observational data, but does not correspond to the boundaries of a really developing cloud, which were, respectively, 80 m and 280 m.

The maximum rate of the longwave radiative cooling ($-2.9° \text{ h}^{-1}$) is observed 60 m below the upper cloud boundary, a maximum of solar heating ($0.65° \text{ h}^{-1}$) is less by a factor of 4.5 and located 15 m lower at a Sun elevation of 23°. Since the cloud is not thick and its bottom is sufficiently low, there is no longwave radiative warming in the lower part. Owing to a shift of maximum radiative cooling to the upper part of the cloud, the air–surface temperature inversion broke and was replaced by the elevated inversion with the bottom at the altitude 300 m. A maximum of the turbulence coefficient ($10.2 \text{ m}^2 \text{ s}^{-1}$) is reached at a height of 100 m.

An interesting effect is observed in the radiation budget. Before the cloud forms at a Sun elevation $\sim20°$, the longwave fluxes are $F_l\uparrow = 276.41 \text{ W m}^{-2}$, $F_l\downarrow = 196.8 \text{ W m}^{-2}$, that is, the longwave balance at this moment is $R_l = -79.57 \text{ W m}^{-2}$, the shortwave balance at this moment is $R_s = 92.13$, that is, the net budget is positive but small, which is explained by the high ice albedo and agrees with the data of radiation measurements.

After the cloudiness forms, $R_l = -7.1 \text{ W m}^{-2}$, and $R_s = 83.76 \text{ W m}^{-2}$, the net positive budget increases, that is, the cloudiness is opaque for longwave and almost transparent for solar radiation, which coincides with the conclusion drawn by Herman and Goody (1976) and leads to a strong greenhouse effect: in the presence of clouds the ice should melt more rapidly. Thus, the low cloudiness can markedly affect the ice regime of the Arctic and is an important climate-forming factor.

The further downward development of the cloud is limited by undersaturation with respect to water near the ice surface, and the upward development is slowed down with the cloud top approaching the top of the boundary layer, that is, a relative stabilization sets in. During 8 hours the top rose 200 m (in fact, by 100 m), the bottom sank 20 m.

The maximum of liquid water content increased and shifted to the top. The temperature lapse rate also increased, especially in the upper part of the cloud, which led to the increase of the turbulence coefficient and appearance of the second maximum near the cloud top.

Most substantial is that the oversaturation of ice is positive only in the upper 100-m cloud layer, where radiative cooling determines the condensation. In the lower 400-m layer, due to the solar heating, the oversaturation is negative, and the droplets evaporate. As a result of evaporation, as over sunlit land covered with snow, the average radius of particles in the bottom part of the cloud increases.

If, as is often done in hydrodynamic models, the bottom of the developing cloudiness is developed from the condition of saturation, its height will be strongly overestimated (in this case the humidity gets saturated 400 m below the bottom), and the radiative heating, both longwave and solar, will happen to be under the cloud. This can be one of the reasons that calculations by Herman and Goody (1976) have given a stratified cloudiness (evaporation of the middle layer). The numerical modelling considered (Marchuk et al., 1986) has shown that at an altitude of 200–300 m the evaporation is most intensive, and with the use of the hydrodynamic approach, the water can completely evaporate here, leading to a stratification. This experiment has not revealed any stratification of clouds, however.

The existence of a two-layer cloudlines, described by Herman and Goody (1976), could have been connected with the proximity of the frontal surface. During the further development of the cloud no qualitative changes take place. Even in 24 hours its boundaries still stay at altitudes 60 m and 750 m.

The principal difference in the dynamics of macro-parameters of both calculated and observed clouds is that, in fact, the rate of the transfer from clear-sky weather to cloudy weather is lower than in the numerical experiment, with a slower increase of the geometrical thickness of the cloud, whose lifetime in the Arctic can be 3 to 5 days. Apparently, this is connected with the fact that in calculations the initial relative humidity, U, has been assumed to be constant with altitude, whereas the observed vertical profile of humidity before the cloud formed had been characterized by decrease with height. In the numerical experiment, with the initial humidity decreasing with altitude, the cloud develops upward much more slowly, and the whole process is closer to the quasi-stationary one. Thus, a rapid decrease of relative humidity with altitude (due to constant low temperatures and attenuation of the turbulent exchange with the surface) is an important feature of the arctic regions, substantially determining the cloud regime in the Arctic.

Tables 2.13 and 2.14 give the results of comparison of the calculated and observed radiation characteristics for an advective fog and stratus cloud, which are compatible both qualitatively and quantitatively (within the error of assessment of the radiative characteristics).

During the transformation of the clear weather and the formation of fog and stratus clouds the rate of radiative temperature change due to both SW (solar) and LW radiation increases, but in the net rate of the radiative temperature change the cooling in the cloud layer of the atmosphere prevails.

In this connection, it is of interest to consider the contribution of various heat flux divergences to the dynamics of the development of low clouds. At initial stages of cloud formation the longwave cooling and negative turbulent influx due to the temperature

Table 2.13. Comparison of calculated (numerator) and observed vertical profiles of radiative characteristics of advective fog at a Sun elevation of 23° (35° in the experiment)

H (km)	F_s (W m^{-2})	ΔF_s	$(\partial T/\partial t)_s$ (°C h^{-1})	$F_l\downarrow$ (W m^{-2})	$F_l\uparrow$ (W m^{-2})	F_l (W m^{-2})	ΔF_l	$(\partial T/\partial t)_s$ (°C h^{-1})
$\dfrac{0.08}{0.07}$	$\dfrac{182}{85}$			$\dfrac{314}{290}$	$\dfrac{300}{280}$	$\dfrac{14}{10}$		
		$\dfrac{21}{19}$	$\dfrac{0.18}{0.18}$				$\dfrac{63}{92}$	$\dfrac{-0.53}{-0.82}$
$\dfrac{0.5}{9.360}$	$\dfrac{203}{104}$			$\dfrac{307}{290}$	$\dfrac{230}{188}$	$\dfrac{77}{102}$		
		$\dfrac{41}{29}$	$\dfrac{0.06}{0.04}$				$\dfrac{48}{-}$	$\dfrac{-0.08}{-}$
$\dfrac{2.85}{-}$	$\dfrac{244}{133}$			$\dfrac{286}{-}$	$\dfrac{61}{-}$	$\dfrac{125}{-}$		
		$\dfrac{56}{21}$	$\dfrac{0.016}{-}$				$\dfrac{35}{-}$	$\dfrac{-0.03}{-}$
$\dfrac{7.2}{-}$	$\dfrac{300}{154}$			$\dfrac{237}{-}$	$\dfrac{77}{-}$	$\dfrac{160}{-}$		

Table 2.14. Comparison of calculation (numerator) and observed vertical profiles of radiative characteristics of a stratus cloud at a Sun elevation of ~32°

H (km)	F_s (W m^{-2})	ΔF_s	$(\partial T/\partial t)_1$ (°C h^{-1})	$F_1\uparrow$ (W m^{-2})	$F_1\uparrow$ (W m^{-2})	F_1 (W m^{-2})	ΔF_1	$(\partial T/\partial t)_s$ (°C h^{-1})
$\dfrac{0.07}{0.06}$	$\dfrac{147}{113}$			$\dfrac{335}{283}$	$\dfrac{328}{278}$	$\dfrac{7}{5}$		
		$\dfrac{27}{50}$	$\dfrac{0.28}{0.45}$				$\dfrac{68}{81}$	$\dfrac{-0.7}{-0.73}$
$\dfrac{0.5}{0.6}$	$\dfrac{174}{163}$			$\dfrac{335}{271}$	$\dfrac{260}{180}$	$\dfrac{75}{91}$		
		$\dfrac{14}{35}$	$\dfrac{0.03}{0.06}$				$\dfrac{43}{-}$	$\dfrac{-0.09}{-}$
2.82	$\dfrac{188}{198}$			$\dfrac{265}{-}$	$\dfrac{147}{-}$	$\dfrac{118}{-}$		
		$\dfrac{68}{21}$	$\dfrac{0.07}{0.02}$				$\dfrac{68}{-}$	$\dfrac{-0.07}{-}$
8.40	$\dfrac{256}{219}$			$\dfrac{202}{-}$	$\dfrac{16}{-}$		$\dfrac{186}{-}$	

inversion is compensated for by 30–40% through latent heat and solar heating, so that the net rate of cooling can reach 1–3°C h^{-1}. With the development of the cloud and increase of the elevation of the Sun the re-arrangement of the temperature stratification leads to a change in the sign of the sensible heat flux divergence, which, together with positive phase transformations and increasing solar heating compensate substantially the LW cooling, so that the net rate of temperature decrease lowers by an order of magnitude and constitutes 0.1–0.3°C h^{-1}. The changes of temperature due to advection and vertical motions are of the same order of magnitude. Thus, the principal conclusion drawn from calculations—and agreeing with observations that reveal the quasi-stationary existence of low clouds—is that a mutual compensation takes place of LW cooling, on the one hand, and positive heat flux divergences due to latent and sensible heat exchange, solar heating, on the other.

A very important effort in the numerical modelling (1-D thermodynamic model) of the impact of clouds on the surface radiation balance (SRB) of the Arctic Ocean has been undertaken by Curry et al. (1993). The most important results of this study have been summarized as follows:

- clouds have a net warming effect on the surface, in contrast to clouds at lower latitudes;
- the annual cycle of cloud-radiative forcing shows warming at all times except for cooling during a few weeks in midsummer;
- for annually averaged cloud fractions less than 0.4, the cloud-radiative forcing is positive (warming) over the entire annual cycle;
- variations in cloud fraction result in variations in the melting period, which may account for observed interannual variability in the melt season;
- low clouds have the greatest impact on the surface radiative flux;
- middle clouds have a smaller impact on the surface radiative flux than low level clouds, but show a larger period and amplitude of summertime net cooling;
- high clouds have relatively small net effect on the surface energy balance, showing a small warming effect over the entire annual cycle;
- cloud optical depth is parameterized to increase with increasing atmospheric temperature, resulting in a net surface warming;
- during summertime, the net flux decreases both for an increase and decrease in drop size;
- the impact of changing ice crystal size is much smaller than that for changing cloud water drop size;
- the modelled equilibrium sea ice thickness decreases with increasing cloud fraction; and
- surface radiative fluxes vary nonlinearly with cloud fraction, due to changes in surface characteristics.

The model experiments by Curry et al. (1993) have provided insight into the nature of two cloud feedback processes in the Arctic. Clouds have been shown to have a net warming effect on the Arctic Ocean surface. This is in contrast to the global average condition, which indicates that clouds have a net cooling effect on the surface. Increasing global temperature is hypothesized to increase the amount of condensed water and thus the cloud optical depth. The results of the calculations for the Arctic showed that increasing atmospheric temperature and thus cloud optical depth results in an increased net surface radiation flux. The so-called "cloud optical depth feedback" is thus a positive climatic feedback mechanism in the Arctic in contrast to the global result obtained by Somerville and Remer (1984). Increasing amounts of pollution aerosol have been hypothesized to decrease water drop size and thus increase cloud optical depth. The results of calculations for the Arctic show that the response of surface radiation fluxes to a decrease in drop size is complex, showing a decrease in the summertime net surface radiation flux and an increase in the wintertime net radiation flux. Overall, arctic clouds do not provide a consistent negative feedback that can be expected to counteract the effects of greenhouse warming.

2.4 ATMOSPHERIC AEROSOLS IN POLAR REGIONS

2.4.1 Introduction

In late October–November the atmospheric circulation forms in the Arctic which hinders the water vapour transport to high latitudes, prevents the transport of industrial wastes together with rapidly moving dry air masses in conditions when the processes of atmospheric cleansing are slowed down. In this period the low-cloud amount, which can cause the washing-out of aerosol from the atmosphere, is small, and the processes of dry deposition on the surface are made difficult due to formation of strong inversions. The inefficiency of the mechanisms of cleansing explains an accumulation of the polluting aerosol components.

The high transparency of the summertime Arctic atmosphere is well known. However, in winter a strong haze is often observed in high latitudes (Arctic Air Pollution, 1986; Barrie, 1986; Barrie et al., 1989a, b, 1989b; Bodhaine, 1989; Bodhaine and Dutton, 1993; Bridgman et al., 1989a, 1989b; Brock et al., 1989; Hansen et al., 1988, 1989; Kondratyev and Binenko, 1981; Kondratyev and Cracknell, 1995; Ottar, 1989; Radke et al., 1984a, 1984b, 1989; Schnell, 1983; Schnell and Raatz, 1984; Shaw, 1982a, 1982b, 1982c, 1985, 1988; Shaw et al., 1993; Thornton et al., 1989; Vinogradova, 1993).

Observations performed in the early 1970s at several arctic stations, revealed much higher turbidity that was expected, and the annual change of transparency was opposite to that recorded in mid-latitudes. The concentration of aerosols in winter turned out to be 10 to 40 times higher than in summer in the North American sector of the Arctic, whereas in the Russian part of the Arctic (Cape Schmidt) this value varies from 5 to 20 times, depending on the concrete meteorological situation.

The airborne measurements near Point Barrow (Alaska) and at Cape Schmidt revealed an increase of the haze concentration with altitude, reaching a maximum at a height of 2–3 km and sometimes higher. Hence, the sources of the haze aerosol had not been local. Studies, started in 1976, revealed an increased concentration of vanadium and manganese due to the industrial pollution; this was a first indication that the arctic haze had been not of natural but of anthropogenic origin. This was also confirmed by the colouration of dust samples (whereas in summer they had been colourless and slightly brown) due to the presence of soot carbon, which could have been the product of incomplete burning.

Organic aerosols constitute about $1 \, \mu m \, m^{-3}$, with surprisingly large amounts of black carbon, 0.3–$0.5 \, \mu m \, m^{-3}$. The total concentration of the submicrometre fraction ($r < 1 \, \mu m$), including SO_4, V, Mn, Pb, Cd, Fe, As, Sb and Se, was $4 \, \mu m \, m^{-3}$.

The most convincing indicator of the anthropogenic origin of haze is a high concentration of sulfates. From the data of observations at Barrow, the sulfates constitute most of the mass of haze particles and, probably, result from oxidation from sulfur dioxide, especially in summer.

The seasonal variation of sulfates and vanadium sampled at Barrow showed that maximum concentrations were observed in March ($2 \, \mu m \, m^{-3}$ and $1 \, ng \, m^{-3}$, respectively); in summer, minimum concentrations were in August and constituted $0.1 \, \mu m \, m^{-3}$ (SO_4) and $0.02 \, ng \, m^{-3}$ (V). On some days of March–April at Barrow and Spitsbergen, concentrations of sulfates of 5 and $10 \, \mu m \, m^{-3}$, respectively, were observed. For comparison, the concentrations of sulfates in the background regions of the northern hemisphere constitute 0.1–$0.2 \, \mu m \, m^{-3}$. The high acidity of precipitation also confirms a great amount of sulfates in the Arctic.

Shaw et al. (1993) have substantiated a model of the polluted arctic troposphere to estimate the magnitude and seasonal variation of the climate-forcing function of arctic haze. For such an assessment they have used three lognormal mass models with properties consistent with the observations in Alaska (Table 2.15) where r_j is the geometric mean radius of the ith mode, σ is the geometric mode standard deviation, and N and M are the particle number and mass concentrations, respectively. The estimates show that the major perturbation to the radiation budget is the lowering of the albedo (heating) of the Earth–atmosphere systems around the vernal equinox and is due to a trace amount (about 5% by mass) of black carbon associated with the removal-resistant submicrometre mode of aerosols. The decrease of the radiation budget at the troposphere level is about $5 \, W \, m^{-2}$. The black carbon over the reflecting polar ice/snow introduces a heating of about 1.5°C per day into the haze layer.

Table 2.15. Parameters for lognormal size distribution

Mode	r (μm)	σ	N (cm^{-3})	M ($\mu g \, m^{-3}$)
Fine	0.02	1.8	400	0.05
Accumulation	0.15	1.4	340	5.0
Coarse	2.5	2.0	0.41	5.0

2.4.2 Observations in the Russian Arctic

During a long period from 1973 onwards, the scientists of the Main Geophysical Observatory and the Institute of Arctic and Antarctic (Russia) have been carrying out ground-based and airborne-based investigations of aerosols in the polar regions. Measurements have been made of the aerosol vertical profiles (up to 8 km) and the size distribution and chemical composition of aerosols (First..., 1981). Simultaneously with aerosol measurements, radiative properties of the atmosphere were measured, including

SW and LW radiation fluxes. Observations were made in the area of the Chukotka Peninsula, in the regions of Anadyra, Cape Schmidt and the Stations Chokurdakh and Tixi.

Morphological analysis of aerosol samples has shown that the lower layers of the troposphere are characterized by the presence of oval particles whose shape is close to a sphere. They are sufficiently dense and look homogeneous. Particles in the samples taken over open waters are especially typical. The particles in samples taken in the middle troposhere were less spherical than those from the lower troposphere. Roughly, 70% of the total number of particles were irregular in shape.

The upper troposphere is characterized by the presence of loose irregular coagulants, particles of the "chain" type, small particles of dense matter. Observations yielded valuable information about the chemical composition of arctic aerosols in regions where measurements have not been made earlier. Table 2.16 presents these data together with the results for other polar regions (First..., 1981).

The amount of soot particles can be considerable from the viewpoint of their effect on the albedo of the snow-ice surface and cloud cover properties. A thorough analysis of the results available is required from the point of view of the arctic haze impact on the regional and global climate. The effect of the aerosol deposition on the ecologically sensitive biomes of the tundra can also be considerable. This has been demonstrated by Adamenko et al. (1991a, 1991b).

Later on, balloon impactor measurements of the vertical structure of atmospheric aerosols have been accomplished by scientists of the St. Petersburg University Institute of Physics and the Voyeykov Main Geophysical Observatory. The impactor used to sample atmospheric aerosols is a continuously moving substratum (formavar film) located under the nozzle, through which an air flow containing aerosol particles is pumped at high speed. Starting from 1987, balloon measurements were based on using a two-cascade impactor. The first cascade was catching gigantic particles, among which there were particles that had possibly been emitted to the atmosphere by the shell and surface of the instruments. Besides, the first cascade protected the substratum of the second one from possible contamination during the impactor's landing.

Basic information concerning the structure and size distribution of atmospheric aerosols has been obtained from electron microscope analysis of the impactor samples. The density of particles whose size exceeded the minimum, d_{min}, has been estimated. The formavar films made it possible to identify particles with $d_{min} > 0.005$ µm with the help of the microscope. In our analysis we restricted the minimum size of particles to 0.01 µm. Within the range of particle sizes up to $d = 0.2$ µm, the uncertainty in estimating the coefficient of efficient trapping of the particle on the substratum is very large, particularly for altitudes $H < 10$ km. Thus, the data for $d < 0.2$ µm can be considerably underestimated.

Morphological analysis makes it possible to identify the following types of particles: particles of mineral origin, particles of organic substance, H_2SO_4 and sulfate particles, conglomerates of smaller particles, particles of the soot fractal type, particles of mixed origin, homogeneous spherical particles of dense matter (high temperature condensate, e.g. micrometeorites), crystals resulting from reactions in the atmosphere (*in situ*). Chemical and elemental analyses were made using the methods of mass spectroscopy, IR

Table 2.16. Elemental composition of polar aerosols (ng m^{-3})

Region	Period of observation	Na	Md	V	Mn	Cu	Pb	Ca	Br	SO$_4$
Greenland	February–April 1978	130	23	0	0.7	0.3	8	—	9.4	—
Novaya Zemlya	April–May 1984	440	300	1.0	6	—	1.3	0.05	12	—
Wrangel Island	May 1984	1180	60	1.0	10	12	1.8	0.06	3.5	520
Alaska	February 1978	310	—	1.5	—	—	—	—	—	2200
Point Barrow	May 1978	—	—	—	—	—	—	—	—	610

spectroscopy, neutron activation and X-ray fluorescence. For this purpose, samples taken from different altitudes within a broad range were used in combination (usually three layers are considered during a flight: 4–10 km, 10–15 km, over 15 km). In some cases, the neutron-activation analysis makes it possible to identify elements with a sufficiently high vertical resolution.

In the cases of observations performed in 1989, in Appatity, all the measured concentrations of chemical compounds and elements turned out to be at the noise level. It is only when IR spectral analysis is employed for altitudes above 15 km that an absorption band appears, determined by the presence of H_2SO_4 in the samples. Its concentration has been estimated at 100–250 μ g/m^{-3}. A neutron-activation analysis has revealed the presence of trace amounts of Fe of about 10–15 μ g/om^{-3} in the lower atmospheric layers.

One can speak with confidence about two layers of increased concentration of aerosol particles in the polar stratosphere: at altitudes 14–16 and 20–21 km, respectively. The data of the electron microscope analysis reveal certain regularities of altitude-dependent variations in both the size distribution and morphological structure of aerosol particles. To increase the reliability of size distribution data, the results obtained for different altitudes were classified into three groups: 5–10, 15–20, and 24–30 km, respectively.

Examination of these data shows that the amount of particles with $d < 0.1$ μm, has been greatly underestimated in the arctic stratosphere, in comparison with the lower atmosphere. The particle size distribution is characteristic of a slow decrease in the concentration of particles with the growth of their size. In the lower stratosphere, there was a large amount of particles in every size range. At altitudes above 24 km, the number of the smallest particles ($d < 0.03$ μm) decreases with altitude. Apparently, there are no sources there to generate particles of these sizes. The maximum particle size distribution is in the range of $d = 0.05 \div 0.15$ μm. This can be accounted for by the large lifetime of such particles in the stratosphere; hence, their growth due to condensation and coagulation.

The regular features of variations in the morphological structure of aerosol particles as a function of altitude reveal themselves very clearly. The general trend is as follows: the troposphere contains sufficiently large particles of mineral origin, conglomerates of smaller particles of organic substance, and sometimes H_2SO_4 particles. About 10% of the total amount may be classified as soot particles. As altitude grows, the proportion of H_2SO_4 particles also increases markedly, reaching 50–60% of the total amount by the altitude of 9–10 km. On the whole, these are spherical liquid particles with individual small solid inclusions (the electron microscope shows traces of dried-up droplets). Even the conglomerates observed of solid particles apparently contain H_2SO_4 (among individual microparticles).

In the lower stratosphere, the proportion of spherical H_2SO_4 particles continues to increase. There are hardly any particles containing organic substance. Particles of mineral (probably, volcanic) origin can be identified. In higher atmospheric layers, particles of irregular shape can hardly be found. Actually, all particles are either H_2SO_4, or sulfate. There are structures representing non-contrasting formations—irregular conglomerates which look like array-webs of fine particles. One's imagination refuses to admit their existence in the atmosphere. It is possible that an array of such small particles was formed out of material sedimented on the film. Data available are insufficient to model optical characteristics of stratospheric aerosols: first of all, there is no information about the soot component. One can only assume that there are soot particles in polar aerosols and that these exist mostly as independent aggregates of small nuclei of the size $d \cong 0.01\ \mu m$. This circumstance reduces the share of radiation absorbed by aerosols, in comparison with the role of soot as a shell on larger particles of a different origin.

Besides, data on the shape and size distribution of irregular particles (plates, chains, cylinders, etc.) should be specified. An assumption can be made that at night these particles have a certain predominant orientation, and if this is so, their substitution by equivalent spheres is impossible.

2.4.3 Origin of the Arctic haze

An appearance after 1976 of a network in the Arctic to sample aerosols has made it possible in 1981 to make the following conclusions: (i) a considerable part of the aerosol haze observed in the wintertime Arctic consists of anthropogenic aerosols which come, apparently, from Eurasia; (ii) the formation of haze starts in late autumn and reaches its maximum in March–April; (iii) the horizontal visibility in the presence of haze decreases to 3–8 km; (iv) the anthropogenic component of haze consists mainly of sulfates (with an average concentration of $2\ \mu g\ m^{-3}$), organic carbon ($1\ \mu g\ m^{-3}$) and soot carbon ($0.3–0.5\ \mu g\ m,^{-3}$); (v) the aerosol that reaches the Arctic overcomes great distances and during its motion the aerosol survives strong ageing.

In March–April 1982, complex aircraft studies were made of aerosol atmospheric chemistry and radiative characteristics of haze, using the flying laboratory WP-3D "Orion" on extended routes (up to transpolar). Visual observations revealed not only the presence of a multi-year haze but also the brown-orange colour of the surface of the central Arctic ice cover. The haze covered the whole polar cap and propagated up to the flight ceiling (8.4 km). Actinometric observations revealed considerable (not less than 10%) absorption of shortwave radiation by haze. The bottom layer of the overcast haze was at a height of 3 km, and above it there were observed broken layers at different altitudes, which looked like bands. The concentration of soot carbon in the haze particles was only by a factor of 3–4 less than in the region of Denver (USA) and only by an order of magnitude less than New York, and by several orders exceeded that observed over the Pacific Ocean.

Atmospheric aerosol can contribute substantially to climatic variations in polar latitudes. Particles, strongly absorbing radiation, can cause the heating of the respective layers of the atmosphere, comparable with its heating due to molecular absorption. The submicrometre particles of atmospheric aerosol, which are cloud condensation nuclei, affect the optical properties and the size distribution of particles and, hence, the radiative characteristics of clouds. The optically active aerosol, falling onto snow or ice, reduces their albedo, which promotes a more rapid thawing of the snow and ice cover of polar regions.

As has been mentioned above, it was discovered in the late 1970s that considerable concentrations of aerosol particles were observed in the Arctic. If the background concentrations are considered to be $0.1–0.2\ \mu g\ m^{-3}$, then in the wintertime Arctic they reach $1–2\ \mu g\ m^{-3}$, that is, the concentration turns out to be comparable with that observed in some industrial regions ($3–6\ \mu g\ m^{-3}$). The seasonal change of the aerosol concentration is clearly expressed: in winter it is an order of magnitude higher than summer. A maximum is observed in February–April, a minimum in June–July.

Based on numerous measurements, carried out at the network of stations, the conclusion has been drawn that the whole Arctic is subject to the gas and aerosol pollution of anthropogenic origin, with local sources contributing least. The aerosol transport to the Arctic takes place mainly from three regions (the northwest coast of the USA, north Europe and southeast Asia). Among the polluting components, the following have been discovered: graphite carbon, sulfates, vanadium, magnesium, organic matter (including polycyclic aromatic hydrocarbons), silicon-containing substances, and condensation nuclei.

The annual course of pollution is explained by the effect of a number of factors: (i) an increased amount of pollutant outbreaks in winter; (ii) an intensified meridional circulation in winter; (iii) a retarded sedimentation of aerosol particles in conditions of the wintertime inversion; (iv) a reduced washing-out of aerosol in clouds and precipitation.

The characteristic time for the transport of polluted air masses from the mid-latitudes to the Arctic at a distance of about 5–10 thousand km constitutes about 20 days. Apparently, a large-scale turbulent diffusion and not a regular advection is the prevailing factor in the transport. If this holds, then the spatial distribution of aerosol in the Arctic should be homogeneous only with characteristic scales more than 2000 km, which exceeds the characteristic size of the arctic cyclones.

The chemical analysis of precipitation and aerosol in the regions of its sources and sinks, a consideration of the ratio of non-soil elements Mn/V (as a product of oil-burning), compared to soil aerosols of the Al type (whose average composition was: Mn-950, V-135, Al-81 ppm^{-1}), and the concentration of the principal circulation processes in the NH atmosphere shows that it is impossible to explain an increase of the Mn/V ratio in the Arctic only through the influence of the North American regions.

2.4.4 Antarctic aerosols

Systematic studies of atmospheric aerosols in the Antarctic started in 1968 by A. I. Voskresensky at Mirny Station, have been continued between 1975 and 1978 at various locations, including the South Pole, with the summertime values exceeding 10 times those in winter, which suggests a substantial impact of meteorological conditions on the concentration of aerosols.

An analysis of aerological soundings and aerosol measurements at the SP plateau has made it possible to draw the following conclusions: (i) the mean total concentration of aerosols varies from 15 particles per cm^3 (in winter) to 100–150 particles per cm^3 (in summer), with particles of 0.02–0.2 µm prevailing; (ii) in the periods of downward atmospheric motion from the middle and upper troposphere, concentrations of 500–1500 particles per cm^3 are observed due to increasing concentrations of particles smaller than 0.01 µm (in many cases the particles smaller than 0.005 µm prevail); (iii) with an advection of moist air masses from the Weddell Sea to the south polar plateau and strong winds near the surface, the concentration of aerosol at the South Pole was about 200–600 particles per cm^3; (iv) the aerosol concentration increases usually on the summertime polar day.

In continuation of studies of the atmospheric aerosol in the Antarctic, the aircraft measurements of aerosol concentrations were performed in November 1977 over the SP plateau near the South Pole using a portable counter of condensation nuclei. A maximum particle concentration was observed in lower, wetter layers of the atmosphere, located over the near-surface inversion as well as in moist air masses over the Ross ice shelf and in the moist air–surface layer of the Ross Sea.

In all cases (even with crossing the tropopause) a comparatively stable decrease of aerosol concentration with altitude was observed. The specific number density (number of particles per 1 g of air) at the South Pole was higher in the 650–450-hPa layer than near the surface. A homogeneous distribution of the aerosol concentration is observed in the 5000–450-hPa layer in the belt 75–90° S. However, above the level of 450 hPa, the concentration decreases weakly with altitude, which can be explained by the effect of a relatively calm and very cold atmosphere above this level.

Such conditions determine the possibility of the formation of ice crystals through homogeneous nucleation and "washing-out" of aerosols through the Brown coagulation of slowly falling particles. The data obtained show that the aerosol is transported with moist air masses from the seas. With these air masses moving over the continental surface, ice crystals form, which fall out onto the surface, removing water and aerosols from the lower layers of the atmosphere. Therefore the air flow from the lower atmosphere to the South Pole turns out to be dry and almost devoid of aerosols.

However, with strong winds ensuring atmospheric mixing, a relatively high concentration of aerosols can be observed near the surface. In the case of strong downward motions, the air masses enriched with submicrometre aerosols can reach the surface from the 550–540-hPa layer. In each case, when the tropopause was crossed by the aircraft, lower values of aerosol concentration were recorded in the stratosphere than near the surface. It follows that only the middle and upper troposphere can be the source of aerosols in the southern polar zone.

The observations considered never revealed concentrations close to those exceeding 10^3 particles per cm^3 near the surface, as has been recorded by earlier observations. Hence, an important problem of the origin of rare aerosol "storms" remains unsolved.

The balloon measurements of aerosol have made it possible to determine the vertical profile of particle concentration with the radius more than 0.15 μm and to reveal the presence of a stratospheric aerosol layer 7 km thick localized at a level of 10 km, on average. The calculated aerosol optical thickness of the atmosphere at a wavelength of 500 nm was 0.0035 ± 0.002, but despite its small value, it constitutes 1/4 of total aerosol optical thickness of the atmosphere over the Antarctic.

As has been mentioned above, a specific feature of the antarctic atmosphere, located far from other continents, is the presence of a considerable concentration of aerosol particles. This testifies to the existence of the long-range transport of particles under the influence of the processes of diffusion and large-scale circulation.

The Antarctic is a unique natural laboratory for studying the transport of particles at long distances and the transformation of minor atmospheric components.

In the troposphere and lower stratosphere over the highly located antarctic ice plateau, submicrometre particles of aerosol are observed whose concentration varies from several particles to thousands of particles per cm^3. Shaw (1988) gave an overview of the results of studies of the origin and properties of the antarctic aerosol, its sources and interaction with the ice cover of the antarctic plateau. In this connection, a model of the diffusional transport of particles (as a problem with boundary conditions) was considered, which assumes the existence of two modes of aerosol (Table 2.17). The smallest particles (Aitken nuclei) consist, apparently, of the transformed products of nucleation of gas components. The age of these particles does not exceed 2–3 days, they form locally over the ice caps at a rate of about 4×10^{21} g cm–3 s^{-1}. The concentration of the Aitken nuclei in the troposphere varies within the range 10^2–10^3 cm^{-3}, except for the lower turbulent boundary layer several hundred metres thick, where the concentration of the smallest particles decreases strongly. Apparently the Aitken particles consist of the transformed gaseous sulfates and hydrocarbons.

Table 2.17. Characteristic values of the parameters of the two-mode model of the antarctic aerosol

Parameter	Mode	
	Large particles	Aitken nuclei
Average radius (μm)	0.14	0.005
Average concentration (cm^{-3})	40	1000
Density of particles (g cm^{-3})	2.16 (NaCl)–1.18 (NH$_4$)$_2$SO$_4$	5.23×10^{-19}
Mass of particles (g)	5.78×10^{-13}	
Average mass concentration (ng m^{-3})	300	0.65
Mixing ratio with respect to air (ppm^{-1})	0.38	0.001
Layer thickness (km)	10	10
Total content in the atmosphere (g cm^{-2})	1.88×10^{-7}	3.9×10^{-10}
Total number (cm^{-2})	5.0×10^5	1.5×10^9
Total optical thickness (at 0.5 μm)	0.013	10^{-8}
Stox rate of sedimentation (cm s^{-1})	5.0×10^{-4}	2.8×10^{-7}
Coefficient of diffusion (cm^2 s^{-1})	2.0×10^{-6}	2.0×10^{-3}

The concentration of larger particles of the second mode, with a radius of about 0.14 μm, constitutes approximately 0.5 cm^{-3}. These particles contribute most to the mass of aerosol and are of great interest, since they, no doubt, come from remote regions and sediment onto the surface, forming the aerosol inclusions to the polar ice. Long-term variations in their concentrations in ice cores can serve as indicators of variations of palaeoclimate.

The sources of the particles of the second mode, by their priority, can be the following: (i) unidentified sulfate sources (the share of sulfate aerosol is 83%); (ii) oceanic sources encircling the Antarctic (about 7%); (iii) arid zones of the southern hemisphere (Australia, and the Kalahari and Atacama deserts), from which the aerosol is transported above the clouds (4%); (iv) extraterrestrial sources (6%); (v) bases on the antarctic continent.

The total concentration of particles on the antarctic air constitutes 170–300 ng m^{-3}, and for various elements: Na 3–70 ng m^{-3}; Al 0.6–0.8 ng m^{-3}; S 60 ng m^{-3} (soil aerosol 7–10 ng m^{-3}; sea salts 10–24 ng m^{-3}; sulfates 150–250 ng m^{-3}; chondrites 0–9 ng m^{-3}); in snow the total mixing ratio is 150–250 ppb, and for various elements: Na 3–20 ppb; Al 0.6–1.2 ppb (soil aerosol 7–15 ppb; sea salts 10–69 ppb; sulfates 60–115 ppb; chondrites 0.12 ppb).

Turbulent diffusion can be considered as a major mechanism for the transport of particles. Estimates of turbulent diffusion confirm the assumption concerning the sources of the particles of the second mode and make it possible to estimate their contributions. The transport of particles from continental sources through the turbulent diffusion prevails, no doubt, in the middle and upper troposphere.

The particles are removed from the free atmosphere over the Antarctic through diffusion to the turbulent boundary layer or as a result of coagulation with cloud particles or precipitation. The major mechanisms for removal are sedimentation onto the plateau surface $(2.0 \times 10^{-14}$ g cm^{-2} s$^{-1})$, snow flakes $(1.6 \times 10^{-14}$ g cm^{-2} s$^{-1})$, ice crystals $(0.2 \times 10^{-14}$ g cm^{-2} s$^{-1})$, and nucleation $(0.6 \times 10^{-14}$ g cm^{-2} s$^{-1})$.

The Aitken nuclei are removed mainly through diffusion to ice crystals $(0.006 \times 10^{-14}$ g cm^{-2} s$^{-1})$, as well as diffusion across the laminar layer to the plateau surface $(0.015 \times 10^{-14}$ g cm^{-2} s$^{-1})$. Numerous mechanisms for the aerosol removal, functioning near the surface, lead to a decrease of the mass concentration of particles in the turbulent boundary layer by a factor of 5–10, which determines the non-representativity of the aerosol samples taken near the surface, from the viewpoint of the properties of aerosols. The aerosol samples taken at altitudes of 2–3 km should be most representative.

Variations in the atmospheric circulation during the last third of the Wisconsin glaciation could have promoted an intensified desertification in mid-latitudes of the southern hemisphere, which had favoured an increased dust-loading of the antarctic atmosphere and a substantial sedimentation of dust onto the surface of the polar ice cap. Since the characteristic time scale of the air-exchange between the southern and northern hemisphere constitutes 1–2 years, note should be taken that the composition of aerosols in both hemispheres is different. This concerns mainly polar regions.

A specific feature of the antarctic atmosphere is its episodical anthropogenic and volcanic pollution. A comparison of the measured optical thickness of the atmosphere (τ_m) along the horizontal and slant paths for the spectral interval 8–10 μm with the results of calculations of radiation attenuation (τ_c) for different models of aerosol consisting of spherical water particles, and for varying meteorological conditions, revealed a sharp decrease of an "excess" optical thickness $\Delta\tau = \tau_m - \tau_c$ with the increasing wavelength. In the interval 8–10 μm the aerosol attenuation decreases with the increasing wavelength. However, the relative contribution of water aerosol to the total optical thickness in the antarctic condition was estimated at 3%. The estimates of the aerosol optical thickness for the atmospheric transparency window 8–10 μm were much below $\Delta\tau$.

Thus, a consideration of attenuation by water aerosol cannot explain the observational data obtained in different conditions. In the spectral interval 8–10 μm with the meteorological visibility exceeding 20–25 km, the contribution of water aerosol to the total attenuation does not exceed 10–20%.

The contribution of aerosol with absorption bands near 9 μm was estimated at 5–13%. Hence, an account of attenuation by both types of aerosol is insufficient to explain the radiation attenuation in the real atmosphere. The absorption by minor gaseous components not taken into account can be one of the possible additional factors.

Kondratyev et al. (1990) have pointed out that in case of undisturbed conditions (in the absence of volcanic eruptions) aerosol optical thickness of the atmosphere in the Arctic is higher than in the Antarctic by 42% and 26% in May and July, respectively.

In a comparison of mean aerosol optical thickness in the Arctic (0.09) and Antarctic (0.013), data on the concentration of aerosols in these regions show that the level of pollution over the north polar region is 7–10 times higher than over the southern one, which is connected with a relative proximity of industrial sources, special features of circulation processes in the atmosphere, geographic and demographic differences between the hemispheres. This has been confirmed by the 54-hour flight (across the North Pole) of a Boeing 747 on 28–31 October 1977, during which measurements were made of the aerosol number density, some gas components of the atmosphere (ozone, carbon monoxide), and meteorological parameters. The flight took place mainly in the 238–162-hPa layer. Measurements of the aerosol concentration were made using an automatic condensation nuclei counter.

At the same time, the US Air Force aircraft LC 130R set its course for the Antarctic, at a lower height, carrying a photoelectric counter, which made it possible to obtain data of aerosol measurements taken at two levels. Measurements were also made on board this aircraft of the number density of large particles greater than 0.45 μm in diameter. The concentration of these particles constitutes less than 1 cm^{-3} in the NH stratosphere and in the eastern sector of the NH troposphere (the flight was made from San Francisco along 120° E to the North Pole, and then along 5° E, landing in London; the flight was continued along 3° E to the equator with the subsequent turning eastward to land in Cape Town; then the aircraft flew along 180° E to the South Pole and along 173° E from the pole to Oakland, New Zealand; the last part of the route was along 156° E across Hawaii to San Francisco).

In the SH stratosphere the aerosol concentration grew gradually, reaching a maximum of 1.6 cm^{-3} in the belt 50–60° S. Close-to-maximum values of concentration were observed in the upper troposphere over the tropical Pacific Ocean and Africa in the regions of the ITCZ.

The latitudinal distribution of the total concentration of the Aitken nuclei is governed by other laws: in the stratosphere of the Arctic and mid-latitude NH the concentration varies within the range 10–40 cm^{-3} and in the respective zones of the southern hemisphere it decreases to 10–25 cm^{-3}. In both polar regions the concentration of the

Aitken nuclei decreases with latitude. The tropical troposphere is characterized by the symmetrical distribution of regions with larger concentrations over land and ocean.

The share of large particles in the SP stratosphere constitutes 6–20%, but in the NP stratosphere it decreases to 5%, and in the maritime near-surface air it falls to 0.1–1%.

A comparison of measurement data obtained from the two aircraft flights reveals, as a rule, a decrease of the aerosol concentration with altitude, but directly over the tropopause (the level 270 hPa), in the belt 50–55° S, the concentrations about 100 cm^{-3} were recorded near the zone of the strongest wind.

An analysis of the measurement results suggests the conclusion that near the jet streams the tropospheric aerosol penetrates the stratosphere, and the near-surface aerosol gets to the free atmosphere mainly in the ITCZ regions and in the regions of the antarctic polar front. This is confirmed, for example, by increasing aerosol concentrations in the belt 4–11° N up to more than 250 cm^{-3} and sometimes to 1500 cm^{-3}. However, large particles are distributed over the troposhere and the stratosphere comparatively homogeneously at all latitudes, with a concentration of about 1.0–0.5 cm^{-3}.

Thus, the data presented suggest the conclusion about the presence of the aerosol haze in polar regions, especially in the northern hemisphere, that there is a still-increasing proportion of anthropogenic and natural aerosols, which can affect the formation of clouds, the radiation regime, and climate changes in the Arctic.

2.4.5 Deposition of heavy metals

Global-scale, long-range transport of aerosols results in the deposition of atmospheric aerosols on the surface all over the globe. Based on observational studies performed at the North Pole, on lakes Ladoga and Onega, in the suburbs of St. Petersburg, and watersheds of the Onega–Ladoga system, and Neva Bay, a quantitative estimate has been made of the fallout of dust and some metals from the atmosphere, most of which are either heavy metals (HM) or trace elements (Adamenko et al., 1991a, 1991b). These assessments have been made using X-ray-fluorescence analysis of the filters (Vladipore 0.45-μm pore diameter membrane filters) on which a solid deposit from snow cover had been filtered.

Table 2.18 shows data on relative concentrations of conditionally insoluble forms of HM, calcium and potassium in snow cover on the ice cover of Ladoga and Onega lakes, at the station North Pole-28 during the annual drift in the northeastern Arctic north of 84° N, as well as relationships between pre-industrial and present-day concentrations of some chemical elements.

Analysis of these data suggests the following:

(1) In the remote regions of the Arctic and northwestern Europe a combination of almost the same chemical elements falls out from the atmosphere—the same elements can be identified whose concentrations in snow cover are from hundredths of a microgram to tens and even hundreds of micrograms per litre of water solution.

Table 2.18. The ratios of concentrations of chemical elements and dust in snow cover on lakes Ladoga (C_L) and Onega (C_O); at the North Pole in winter (C_{mw}) and summer (C_{ns}); and preindustrial (C_p) and current (C_c) concentrations in the northwestern part of Russia

Element	C_L/C_O	C_L/C_{ns}	C_L/C_{nw}	C_O/C_{nc}	C_O/C_{nw}	C_c/C_p
K	2.9	12.1	—	4.2	—	—
Ca	1.1	5.6	—	5.0	—	9.8
Ti	3.4	19.6	9.2	5.8	2.8	—
V	3.0	6.0	2.0	2.0	0.7	—
C	1.1	6.2	1.9	5.4	1.6	—
Mn	1.6	25.0	8.5	15.0	5.2	—
Fe	1.8	27.0	15.8	17.7	8.6	10.6
Ni	0.9	5.2	3.0	6.1	3.5	—
Cu	1.3	9.1	7.8	7.1	6.1	18.3
Rb	1.2	2.7	3.3	2.1	2.6	15.4
Pb	2.7	21.8	—	8.2	—	—
Sr	1.6	26.3	10.2	16.8	6.5	—
Zr	1.5	45.5	14.3	29.7	9.3	—
Br	—	—	—	—	—	—
Zn	1.3	10.7	8.2	8.1	6.3	20.4
Dust	1.9	12.6	6.8	6.5	3.5	—

(2) These elements are located, by order of priority, in the following successions:
  - Ladoga (concentrations in $\mu g \, l^{-1}$ are given in parentheses): Fe (155), K (82), TI (17), Ca (13), Mn (2), Zn (2), V (1), Cr (0.8), Zr (0.7), Cu (0.6), Pb (0.6), Sr (0.4), Ni (0.2), Rb (0.2), Br (0.1), dust (2.6 mg l^{-1}).
  - Onega (in $\mu g \, l^{-1}$): Fe (84), K (29), Ca (11), Ti (5), Mn (1), Zn (1), Cr (0.7), Pb (0.5), V (0.4), Zr (0.4), Ni (0.3), Sr (0.3), Rb (0.1), Br (0.1), dust (1.36 mg l^{-1}).
  - North Pole (summer, $\mu g \, l^{-1}$): K (7), Fe (6), Ca (2), Ti (0.9), Pb (0.2), V (0.2), Zn (0.1), Cr (0.1), Mn (0.99), Cu (0.07), Ni (0.05), Br (0.020), Sr (0.017), Zr (0.015), Rb (0.010), dust (0.21 mg l^{-1}).

(3) The wintertime deposition in the Arctic exceeds 2–3-fold that in the summer, which is explained both by the winter duration and by differences in atmospheric stratification, which is more stable in winter than in summer.

(4) For almost all elements, the deposition on Ladoga is 20–40% greater than on Onega, which is explained by the proximity of Ladoga to relative large sources of atmospheric pollution, compared to Onega, as well as by prevailing winds with southern or western components in northeastern Russia in cold seasons.

(5) Analysis of the available data on the fallout of chemical elements from the atmosphere in cities with multi-million populations and with diversified industry suggests that in such cities the deposition of some chemical elements exceeds by one to two orders of magnitude the fallout for the remote regions in the northwest (Ladoga and

Onega) and exceeds by three to four orders of magnitude the input of metals at the North Pole. The latter data can be considered a measure of the HM deposition for the global background conditions of the northern hemisphere.

(6)   The HM deposition on Ladoga is 3–45 times stronger than at the North Pole in summer and 2–16 times more intensive than in the central Arctic in winter.

(7)   The background deposition of HM on relatively pure Onega is 2–30 times more intensive than at the summertime North Pole and 2–9 times stronger than in the central wintertime Arctic.

(8)   The present deposition of such elements as Ca, Fe, Cu, Pb, and Zn exceeds by 10–20 times that of 150–200 years ago.

(9)   Differences between the background hemispheric (North Pole) values of dust concentrations and the background regional (Ladoga) values reach one order of magnitude, and for Onega are half as much.

(10) The concentration differences of the deposited Hg are, on the average, the same as in the case of total dust. However, the concentrations of lithophyll metals (metals with low enrichment coefficients in particles of atmospheric aerosols) (manganese, strontium, iron) are tens of times smaller in the arctic dust, whereas the concentration of atmophyll metals (metals with high enrichment coefficients in particles of atmospheric aerosols), for which the industrial contributions are significant, differ less.

(11) The atmospheric flux of lithophyll elements (calcium, magnesium, iron) during the industrial epoch has increased by a factor of 5–11, and that of atmophyll elements (lead, zinc, copper) by a factor of 15–20. This points to the fact that the present background level of atmospheric deposition assessed for the hemisphere from snow samples at the North Pole are of about the same order of magnitude as in the mid-latitude northwest in the pre-industrial epoch.

3

Remote sensing of polar ice and snow properties for the purpose of climatic monitoring

3.1 REQUIREMENTS FOR THE MULTI-SPECTRAL REMOTE SENSING OF ICE AND SNOW PARAMETERS

The space-borne remote sounding techniques are the key means of studying climate. In the polar regions of the Earth, in conditions of the persistent existence of the drifting snow-ice cover and the presence of continental ice, it is particularly difficult to organize routine hydrometeorological and oceanographic observations, so the development of techniques and means of remote sensing of the most important climate characteristics (such as: the distribution of pack ice by its size, age, thickness and hummocking; the concentration and dynamics of motion; processes of fast and drift ice formation, freezing and melting; transformation of heat fluxes through snow and ice; the state of the arctic atmosphere, etc.) is a very important and urgent problem. It is also important to point out that it is very difficult to obtain long-term data sets of the sea ice characteristics that are required to study climate.

A significant aspect of the monitoring of the polar regions should be the comprehensive ground-, air-, and space-based survey of the snow and ice cover. Barber et al. (1992) compiled tables characterizing the problems and requirements for the ice cover monitoring in the polar regions. Table 3.1 gives the general information which serves the basis for the development of instrumental complexes for remote sensing of the parameters enumerated in the table. This table, of course, should be considered tentative, but without doubt these requirements can help to assess the present-day possibilities of obtaining multi-spectral remote sensing data and of evaluating further development of satellite remote sensing techniques.

Table 3.1. Glaciological aspects of the ecomonitoring of polar regions and possibilities for remote diagnostics

Parameter	Objectives	Spatial resolution	Temporal resolution	Remark
1	2	3	4	5
Ice cover concentration	Ice thickness to be taken into account in the models of dynamics. Ship steering and solution of tactical problems in navigation	Acceptable > 100 m; pixel 10 m; 100 m; statistical distribution of thickness within the swath width > 1 km	6 hours (operatively) 1 month (regional statistics)	
	Navigation strategy	1 km	6–24 hours	Fast data output (< 3 h)
	Heat and mass exchange between the atmosphere and the ocean; climate change index	1 km (lower limit); statistical distribution	3 days	
	Distribution of mammals	10 m (lower limit), 1×10^4 m^2 is also useful	1–7 days	
	Estimates of insolation analysis of conditions of photosynthesis in water	1 km^2	1–7 days	Statistical estimates of condensation and type of ice over large areas (10^2–10^5 km^2) are needed

Table 3.1. (*continued*)

Parameter	Objectives	Spatial resolution	Temporal resolution	Remark
1	2	3	4	5
Size distribution of ice	Ice thickness to take into account in the models of dynamics; analysis of wave attenuation	1 m and less	6 hours (operatively) 1 month (regional statistics	
	Tactics and strategy of navigation	Pixel 1–10 m; swath width > 1 km	6-24 hours	Fast data output (< 3 h)
	Energy-exchange (open/ ice-covered water ratio)	1 km (lower limit); statistical distribution	3 days	
	Locations of mammals	10–100 m; swath 1×10^4 m^2 is also width > 10 km	Daily	Detection of small ice-floes with seals is important
Ice cover thickness	Dynamic models	10-m pixels within 100 km; 10 cm thick for marine operations	6 hours (operatively), 1 month (regional statistics)	Combination with data on the size distribution of ice-floes is important
	Tactics and strategy of navigation	2–10-m pixels	1–14 days	Data output during 12 to 24 hours

Table 3.1. (continued)

Parameter	Objectives	Spatial resolution	Temporal resolution	Remark
1	2	3	4	5
	Turbulent heat-exchange; temperature lapse rate in the ice cover	10 km (horizontally); 5 cm (vertical)	3 days	Climate monitoring; trends variability
	Locations of mammals	Horizontal—to detect leads $(1 \times 10^4$ $m^2)$; vertical—±10%	1–7 days	Combination with data on snow cover thickness
Snow cover thickness	Estimates of insolation and analysis of conditions of photosynthesis of algae in the ice cover	1 m (horizontal); ± 10% (vertical)	7 days	Same
	Estimates of ice properties, temperature, etc.; loads on ships	5 km at a distance of 100 km	7 days (for navigation purposes)	Sufficient to meet the requirements of oceanology and biology
	Ship steering	0–2 m; error ± 25%		
	Heat balance modelling	0.2 m; ± 10%	1–14 days	Data output during 12 to 24 hours
	Turbulent heat-exchange temperature profile	Resolution of the vertical stratified structure		Synoptic processes in the Arctic; distribution of the thawing ice spots

Table 3.1. (*continued*)

Parameter	Objectives	Spatial resolution	Temporal resolution	Remark
1	2	3	4	5
	Ice algae: light transmission	0.1 m (horizontal "point" observations); 1 km (horizontal distribution)		Time resolution should increase with temperature growth
	Distribution of seals' rookeries	± 10% (horizontal); 1–10 m (point-by-point); several km^2 (survey)		Important data on diurnal change and snow transport
Age characteristics of ice	Input information for dynamic models	3 classes (1-, 2-, and multi-year ice); 10-m pixels with swath width > 1 km	1 day	Data output during 12 to 24 hours (1-year ice)
	Thermodynamic properties and indicators of climate (with respect to ice ages)			Ratio of 1-, 2-, and multi-year ice; indicators of the dynamics of the atmosphere–ocean heat exchange
Ice density	Snow supply modelling	10% of observed density	1–14 days	Daily mapping

Table 3.1. (*continued*)

Parameter	Objectives	Spatial resolution	Temporal resolution	Remark
1	2	3	4	5
Ice cover roughness	Electromagnetic interaction	Error ± mm	Hourly	
	Momentum exchange	100 m (horizontal); 10 cm (vertical)	3 days	
	Turbulent fluxes	10 m–1 km (horizontal); 1 cm (vertical)	Hourly	
Wind speed	Ice motion forecast	Winds at a height of 10 m over the grid with a step of 5–10 km; error 1–2 m s^{-1}	6–24 hours	Fast (1–3 h and faster) data output
	Energy exchange with the atmosphere	Winds at height of 10 m; error 5 m s^{-1}	3 days	
Ice cover topography	Marine operations: loads on constructions	*Investigations*: geometry of ice hummocks (height ± 25 cm; width ± 50 cm; depth ± 50 cm; porosity ± 0.05; length, orientation)	24 hours–1 month (regional statistics)	

Table 3.1. (*continued*)

Parameter	Objectives	Spatial resolution	Temporal resolution	Remark
1	2	3	4	5
		Operatively: geometry of ice hummocks (height ± 0.5 m; width ± 1 m; keel ± 0.5 m; length, orientation)		
	Surface deformation	Length, orientation; height ± 1 m; keel ± 1 m	6 hours	
	Ship steering (hummocking) modelling (wind shear near the surface)	± 25%	1–14 days	Data on the ice hummocks distribution statistics (coverage, error ± 15%); width of ice hummocks 5 m (± 15%)
	Convergence/divergence of ice	1–10 km (statistical distribution of meso-scale orientation)	7 days	Local dynamics of ice cover
	Vertical ice cover roughness	1 m (horizontal; point-by-point); several km (survey)	24 hours (pack ice) 7 days–1 month	Cod population in the Arctic; important data on ice keels

Table 3.1. (*continued*)

Parameter	Objectives	Spatial resolution	Temporal resolution	Remark
1	2	3	4	5
	Vertical roughness that controls the snow cover thickness (ecology of seals)	± 10% (vertical); 1–10 m (horizontal, point-to-point); several km² (survey)	7 days–1 month	
Temperature at water–ice–atmosphere interfaces	Indicator of physical and mechanic properties	*Vertical:* deformed ice (50 cm); near the surface of non-deformed ice (1 cm) *Horizontal:* deformed ice (5 m at 100 m); non-deformed ice (10 m–100 1 km at 100 m–100 km for studies and operations respectively)		
	Growth and disintegration of ice cover, heat balance, freezing/melting	Air temperature (1°C); ice temperature profile (0.1°C over the grid 5–10 km)	3–6 hours	

Table 3.1. (*continued*)

Parameter	Objectives	Spatial resolution	Temporal resolution	Remark
1	2	3	4	5
	Sensible heat flux; LW balance, near-surface humidity	0.5°C	15 minutes, 6 hours (synoptic analysis)	Thermo-inertial role of ice cover
	Convection of nutrients near the ocean surface	0.1–1 km 1–10 km	1 hour 24 hours	
	Specific phenomena (upwelling, fronts, etc.)	0.1–1 km 1–10 km	1 hour 24 hours	
Surface radiation budget	Upward thermal emission, albedo, transmission as a function of ice cover characteristics	As a function of surface inhomogeneity	15 minutes (diurnal course), 24 hours (annual course)	
	Effect of thawing, pollution, etc.	Ditto (error in radiation fluxes 0.5–1 W m^2)	Ditto	
	Atmospheric counter-emission	Ditto	Hourly	
	Photosynthetic curve radiation	Survey	Hourly	

Table 3.1. (*continued*)

Parameter	Objectives	Spatial resolution	Temporal resolution	Remark
1	2	3	4	5
Ice cover kinematics	Data on tensions; verification of dynamic models	Pixel 10–100 m for swath width 10 km	1–24 hours	Spatial resolution depends on special phenomena: convergence, shiftings, etc.
	Navigation regions (with respect to convergence/divergence)	Identification of 3 classes of ice—1 km (horizontal); speed cm s^{-1}	1–3 days	
	Cycles and drift of ice to map the atmosphere–ocean interaction	Drifting for 10 km during 24 hours	24 hours	Field of motions: convergence/divergence
Water vapour content in the atmosphere	Latent heat fluxes over ice and open water; cloud distribution	Microscale and mesoscale (up to 10 km)	15 minutes–6 hours	Important estimates of water vapour input from leads
	Convection of nutrients	Survey regime	Daily survey	
Extreme states of ice cover	Dangers for navigation	Pixel < 10 m for swath width > 100 km	1 hour–48 hours	Important consideration of icebergs

Table 3.1. (*continued*)

Parameter	Objectives	Spatial resolution	Temporal resolution	Remark
1	2	3	4	5
	Drifting of harmful formations, interaction with roughness at the ice edge	10 m–10 km (presence/absence)	As a function of concrete formations	
	Flow of icebergs as a climate change index	Pixels < 10 m for swath width > 100 km	Seasonal monitoring every 3–4 days	
	Wintertime distribution of walruses and seals	0.01 km^2	Survey regime	Only regions of broken ice
Oil slicks	Detection of ecological dangers; maintaining of cleaning operations	100 m (horizontal at a distance of 10 km); 2 cm (edge thickness)		Important detection of oil pollution among ice-floes
Parameters of roughness	Ice cover destruction	5 m (horizontal) for 500 m to 100 km	5–24 hours, survey regime	Maintenance of navigation and drilling operations
	Detection of biological phenomena	0.1 m (vertical), survey regime		

Table 3.1. (*continued*)

Parameter	Objectives	Spatial resolution	Temporal resolution	Remark
1	2	3	4	5
Distribution of algae	Thermodynamic models	From 10 m pixels (at a distance of 100 km) to > 100 m	7 days	
	Ice-free phytoplankton	0.1 km^2 (point-by-point) to 10 km^2 (survey)		Important data on reflectance spectra
	Algae in the ice (lower layer of 1-year ice 1 cm thick)	Survey regime with needed horizontal resolution to 0.1 m		Direct detection is possible
	Suspended matter	10–100 m (point-by-point); 0.1 km^2 (survey)		
Ice/open water ratio	Leads: navigation	Length (± km); width (± m); extent (± 10%)	6–24 hours	
	Polynyas: navigation	Size, location (within tens of metres)	7 days	

Table 3.1. (*continued*)

Parameter	Objectives	Spatial resolution	Temporal resolution	Remark
1	2	3	4	5
	Breaks (polynyas as locations of seal population; migration of whales)	0.5 m (width) by kilometres (length)	Survey regime	Very important
	Ice-covered breaks and leads as places for whales			Very important
	Polynyas	± 100 m^2	Ditto	Ditto
	"Windows" for seals	$(1–100)$ m^2	Ditto	Ditto
	Multi-year locations of seal population	Icebergs		Not important
	Ice hummocks and thawed patches	$1–10$ m^2	Ditto	Ditto
Size distribution of ice cover	Radiation penetrating through the ice, concentration, etc.	Details of 1 μm to 100 cm in size; horizontal resolution 10–100 m	Monthly (in winter); daily (seasonal boundaries)	

Table 3.1. (*continued*)

Parameter	Objectives	Spatial resolution	Temporal resolution	Remark
1	2	3	4	5
	Propagation of heat through ice	Millimetres	7 days	
	Electromagnetic phenomena	Ditto	1–24 hours	
Size distribution of snow cover	Heat flux through the snow; radiation	Details from mm to cm; vertical: cm; horizontal: 10 m	Monthly (in winter); daily (seasonal boundaries)	
	Heat exchange through the snow cover	Millimetres	7 days	
	Electromagnetic phenomena	Ditto	1–24 hours	

3.2 VISIBLE AND INFRARED SATELLITE REMOTE SENSING OF ICE AND SNOW AS COMPONENTS OF THE EARTH'S CLIMATIC SYSTEM

The remote sounding techniques in the visible and IR spectral regions to obtain information on the spatial distribution of ice and snow properties have been applied from the very beginning of satellite meteorology and are being operationally used for various practical purposes. These techniques, however, are substantially limited by insufficient illumination of polar regions, as well as by controversial approaches to interpreting data on the reflectivity and emissivity of natural formations; by dependence of the measurement results on viewing geometry and sounding conditions (Sun elevation, cloudiness); and by the need for reliable solution of the atmospheric correction problem.

Therefore, we shall only refer to some principal results in the field of analysis and interpretation of visible and infrared images.

Initially satellite ice data have been obtained for limited regions often with a resolution allowing the identification and tracking of individual ice floes (Parkinson and Gloersen, 1993).

Such data may be useful for various applications study, for example, for ice navigation, for geomonitoring and for protection of the marine environment. These data are valuable for solving local polar seas problems (determination of ice-floe shapes and sizes and calculation of ice velocities) as well as for studying large-scale oceanic processes, including the discovery of the Atlantic "trace" in the Arctic ocean and the indication of vast areas of fresh water propagation into the Arctic shelf seas (e.g. DeRycke, 1973; Dey et al., 1979; Dey, 1981; Ito and Muller, 1982).

A few decades ago only ground-based observations enabled the study of snow cover dynamics in polar regions. Snow parameters were measured only at the few polar hydrometeorological stations. Since 1966, continental snow cover extent has been monitored on a weekly basis using NOAA satellite instrumentation in the visible and near infrared ranges, but the problem of obtaining quantitative and representative information concerning the water equivalent of snow, the snow depth and the snow extent was not solved (Foster and Chang, 1993).

The monitoring of the natural environment is undertaken to forecast variability the climate-forming factors, as well as to substantiate recommendations on the optimization of conditions to preserve the natural environment. In accordance with the scales of observed changes, there are three levels of monitoring: global, regional, and local. For polar regions, the global-scale observations are accomplished to get information on the sea state, sea ice dynamics and continental ice formation, as well as snow cover characteristics (both on sea and on land). The regional-scale observations have the purposes of studying the ice regime of fresh-water lakes and mouth-areas of large rivers, the dynamics of frozen soils and multi-year ice cover, the intensity of descents of snow avalanches, etc. On the local scale, snow supplies in river basins, the state of the water–land interface, various constituents of the water–ice balance and the state of pulsating glaciers, etc., are observed (Kotliakov, 1968).

The large-scale effect of snow cover on climate was first demonstrated in Russia by Voeikov (1889); the importance of studying the trends of the ice cover variability to analyse climatic changes has been emphasized by Brooks (1952); studies by Gernet

(1930) of ice formations have become classical in the USSR. Along with the global factors the account of regional factors also remains important, for example, studies of the regional climatic effect of the ice cover on some polar seas, started by Nansen, has been substantially developed later.

In studies of snow and ice cover, of interest are satellites both in polar and geostationary orbits. The satellites of the first type are the NOAA meteorological satellites of the third generation functioning in almost-polar, Sun-synchronous orbits about 1500 km high, which provides global images in the visible once a day and in the infrared twice a day.

From the viewpoint of hydrological applications, of most interest is the use of the AVHRR data. The Meteor, Resurs, Landsat and SPOT data are also useful (especially for polar regions), though their use is hindered by a rare repeatability of data. The NOAA, Meteor and Resurs satellite data are being used by the Arctic Ice Service of Russia for preparing comprehensive ice maps. These maps are produced systematically three times per week and distributed to icebreakers and ships which sail along the North-East Passage. Such maps can be used for climate studies as well. Figure 3.1 demonstrates an example of the comprehensive ice map for the autumn season (1–7 September 1993). As has been mentioned above, during the period of summer arctic navigation when ice and snow melts, the visible and infrared data permit the assessment of the sea ice concentration only, while to evaluate ice age parameters radar and microwave data are necessary.

The importance of geostationary satellites consists in the possibility of their application for the continuous monitoring of the dynamics of natural processes and for collecting and transmitting data from land and ocean (moored or drifting) automatic stations as well as from balloons drifting in the atmosphere. The impossibility of covering polar areas is a very serious limitation of the geosynchronous satellite data.

As Wiesnet (1974) has shown long ago, the interpretation of the AVHRR data makes it possible to map the snow cover distribution over the areas of more than 5000 km^2 with an accuracy of about 5%. The experience gained from mapping the snow cover in river water basins has also been positive. Of great importance in this case is the use of the Landsat, SPOT and ERS-1 data for calibration purposes. A deficiency of the ERS-1 data is the limitation of the observing area—the swath width for such satellite data is only 100 km.

The spatial resolution of the AVHRR images is high enough to enable one to substantially improve the interpretation of images for assessing the thickness and extent of snow cover. McGinnis et al. (1975) considered an example of such an interpretation using the NOAA-2 data after heavy snowfalls.

A comparison of snow-cover maps based on analysis of images in the visible and from ground data, revealed a good agreement. The location of the snow-cover edge can be determined from satellite data to an accuracy of at least 10 km. The field of brightness for the interval 0.6–7.0 μm visualized in false colours revealed the possibility of an approximate estimation of the snow-cover thickness from brightness data within three categories: below 8 cm, 8–30 cm, and more than 30 cm. The use of quantitative but not calibrated data of brightness measurements with maximum brightnesses chosen for 32 by 32 km^2 squares enabled one to obtain a sufficiently high correlation between brightness and

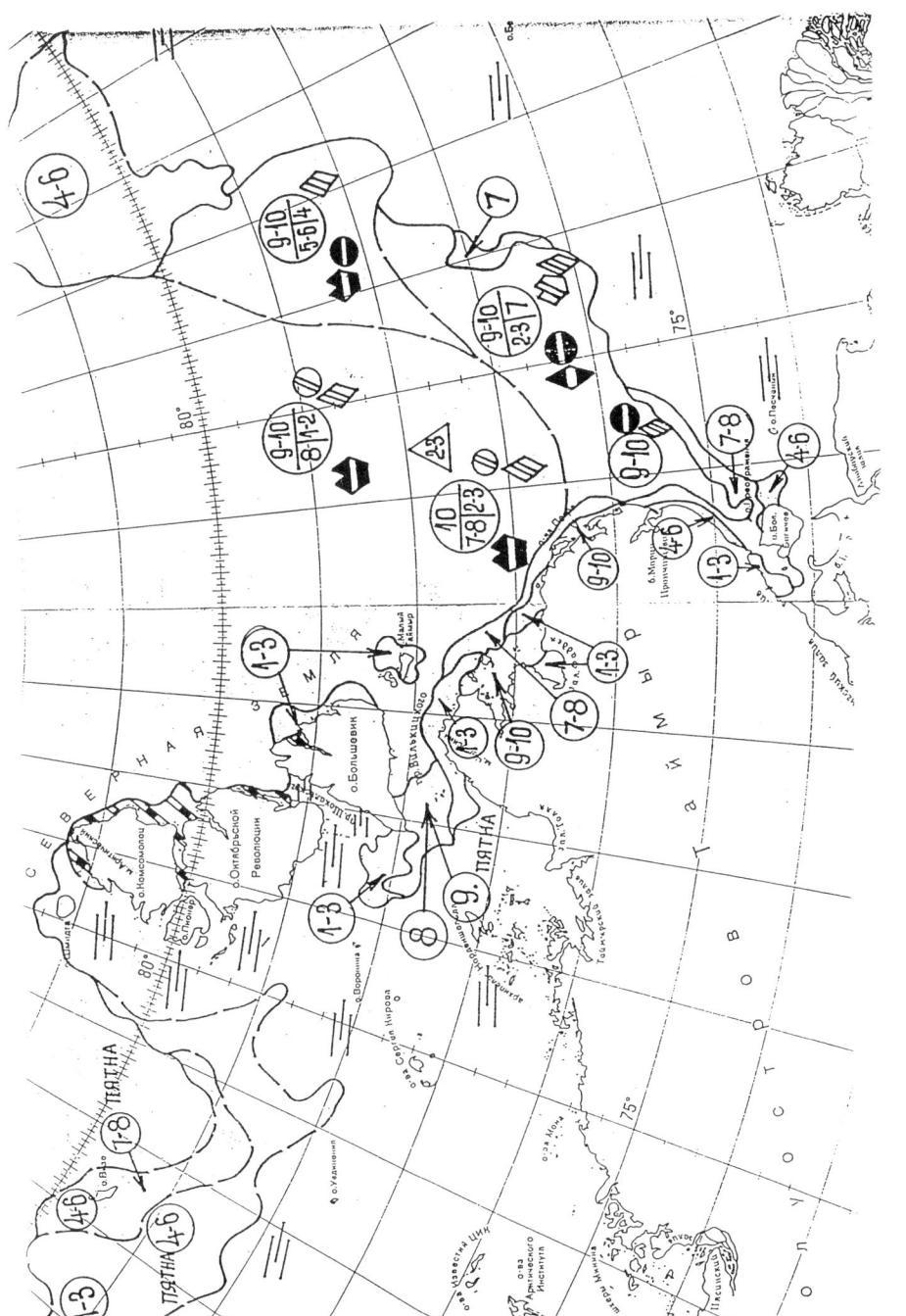

Fig. 3.1. Comprehensive ice map (1–7 September 1993). Data from the Russian Marine Operative Headquarters (Dikson)

thickness of snow cover. With the obtained dependence approximated by parabola the coefficient is 0.84.

Takashima and Morrissey (1976) illustrated the interpretation of the NOAA-3 data for the regions of Baffin Bay and the Barrow Strait in North Canada. In both cases the images enable one to distinguish between eight levels of brightness or surface tempera-ture. The site in the centre of image has the level of brightness 150 (in conditional units). In this case the ice fields are in the range of brightnesses 140–155. The upper cloudiness over the ice cover was excluded with the help of an infrared image in which it is easily recognized as a bright zone with a characteristic texture. The sites corresponding to pack ice in the open sea or to large leads can be easily recognized. Areas of open water and polynyas are seen.

Thus, the examples of earlier studies considered testify to the possibility of recognition of drifting and pack ice, upper and lower clouds, ridged zones, leads, polynyas and other important climatic characteristics.

A substantial contribution of the Landsat data to the monitoring of the ecological situation in polar and circumpolar regions has to be mentioned. For example, an analysis of Landsat-1 and Meteor-3 images of the regions of oil extraction in Alaska and West Siberia, respectively, reveals a strong effect of pipelines, roads, and zones polluted with oil products, on the state of snow cover, on the time of river break-up, etc.

Turner (International Conference..., 1991) processed data of satellite remote sounding for the purposes of meteorological sounding (NOAA TOVS instrumentation) and showed that the errors in retrieving the vertical temperature profiles and water content of the atmosphere in polar regions remained too large, preventing reliable retrieval of the long-term trends. So, for example, the errors in temperature retrieval, caused by insufficiently reliable filtering out of the effect of clouds, constitute about 2–3°C, considerably increas-ing in the atmospheric boundary layer and near the tropopause.

Fletcher (International Conference..., 1991) undertook an analysis of the global data-base of conventional meteorological observations (COADS), which led to the conclusion about the long-term variability of the AGC (Atmospheric General Circulation) character-istics in the polar regions which did not agree with the known results of numerical climate modelling. So, for example, it was found out that the observed global winds enhancement (this refers also to trade-winds and west–east transport) by about 29% in the 1930s compared to the 1920–30s, does not agree with that calculated. During the last century three to four sudden changes of AGC took place. The long-term oscillations of two types have been observed: about 3–4 years (this is connected with the ENSO events), and of the order of 1.3 centuries (in this case the forcing appears, probably, in high latitudes in winter, more strongly manifesting in the Arctic than in the Antarctic). The solar activity can be supposed to be a significant factor of long-term oscillations. Royer (International Conference..., 1991) discovered an 13.6 year cyclicity of sea surface tem-perature (SST) and SAT, which could be ascribed to the effect of lunar tides.

Having processed the cloud data from the Russian meteorological satellites for the period 1971–1985, Makhov (International Conference..., 1991) discovered a general trend towards an enhancement of cloudiness all over the world, as well as in the northern and southern polar regions as the SAT increases, but against this trend, in some regions there was observed an anticorrelation of SAT and cloud amount, especially in high latitudes of the northern hemisphere in winter and in spring. Stone and Cahalan (International Conference..., 1991) note that the intensity of surface air inversions could be a good indicator of climate change in the zone of the Antarctic plateau. Bromwich (International Conference..., 1991) emphasized the fact that precipitation in the regions of Greenland and the Antarctic could be an important factor in the world ocean level variations.

The development of remote sounding of the ice cover and surfaces of the arctic seas is more and more determined by the use of the data of passive and active radar techniques. This is connected with both a large information content of radio-range data and the impact of cloudiness prevailing in polar regions. The basic instruments used in all-weather diagnostics are the following: radio-altimeters, synthetic-aperture radars, scatterometers, and microwave radiometers which since the 1960–70s have been launched on board American, European and Russian satellites. Below the physical principals of the techniques of radio-sounding of the ice and snow covers and frozen soils, as well as their potential for studies of the problem of the Arctic climate, will be considered.

4

The physical basis for microwave sounding of the Earth–atmosphere system with negative air temperatures

As mentioned above, radar and microwave sounding is a general technique of aerospace monitoring of ice and snow climatic parameters. However, in spite of the significant progress in the thematic interpretation of satellite microwave data the accuracy of retrieving the important parameters of ice and snow cover is not sufficient to be used in numerical modelling of Arctic and Antarctic climate. Ambiguous interpretation of the microwave satellite data is determined partly by insufficient knowledge of electrophysical parameters of water, fresh and sea ice, snow cover and frozen soil. Results of Western fundamental theoretical studies on electrophysical parameters of water, snow, and ice, including the sea ice microwave measurements and models, as summarized by Wadhams (1992a), and snow microwave research, summarized by Jones (1983), and by Singh and Srivastav (1993), are well known.

In the chapters below we present the key results of extended theoretical and experimental research, produced in Russia and used by Russian experts but largely not published in the Western scientific journals. This material could essentially broaden common knowledge of microwave emissivity and reflectance properties of natural surfaces at negative temperatures and aid the multi-spectral satellite data interpretation. Chapters 4 and 5 present theoretical considerations regarding the electrical and physical parameters of the cryosphere. Chapter 7 presents the contributions of major Russian programmes to study the Arctic environment using microwave remote sensing and *in situ* data. The key contributions of the major Western remote sensing and *in situ* programmes to studies of the Arctic (for example Parkinson and Cavalieri (1989), Parkinson (1991) and Gloersen et al. (1992)) are presented in Chapter 6.

4.1 ELECTROPHYSICAL PARAMETERS OF FRESH AND SEA WATERS, FLOATING ICE, SNOW, AND FROZEN SOILS—PROBLEMS OF CORRECT EVALUATION

The idea of using the microwave means of obtaining data on the global environment components, including processes occurring in polar regions, was originated in Russia in the mid-1960s soon after the launching of the first Earth research satellite (1957). Studies on the microwave properties of natural surfaces were begun at the same time. The research has included theoretical and observational studies for application in meteorology, oceanography, and terrestrial studies.

The microwave emission of various natural materials and structures is the thermal emission of heated bodies. The physical mechanism for thermal emission consists in the conversion of internal thermal energy into the energy of the electromagnetic field and propagation beyond the heated body. This phenomenon depends on the physical state of the emitting body (solid body, liquid, gas) and is determined from its temperature and emission coefficient (emissivity), varying much in the microwave part of the radiofrequency range, depending on the state of the "emitting body–environment" interface.

The principal difficulty in the interpretation of the information from microwave sensors carried by aerospace platforms consists in a variety of factors differently affecting the radio-brightness temperature T_B of the Earth–atmosphere system. This circumstance, as well as an ambiguous interrelationship between the parameters of the state of the ocean and the atmosphere, measured with conventional *in situ* sensors and retrieved on the basis of remote sensing, have long hindered the use of microwave sounding techniques in hydrometeorological and oceanographic observations.

The radio-brightness temperature of an object $T_{B,\lambda}$ is physically the temperature of a blackbody (BB), whose brightness at a given wavelength of sounding λ is equal to that observed. For the cases of sounding from aircraft, research vessel (RV), or for model-simulation experiments, it is written as

$$T_{B,\lambda} = \sum T_S\, e^{-\sec\Theta \int_0^H \alpha\,dz} + \sec\Theta \int_0^H T(z)\alpha\, e^{-\sec\int_z^H \alpha\,dz}\,dz$$

$$+ \left\{ \int_0^{2\pi} d\varphi' \int_0^{\pi/2} r(\Theta_1,\varphi_1,\Theta',\varphi') \sin\Theta' \cos\Theta'\,d\Theta \right\} \qquad (4.1)$$

$$\times \left[\sec\Theta' \int_0^\infty T(z)\alpha\, e^{-\sec\Theta' \int_0^z \alpha\,dz}\,dz \right] e^{-\sec\Theta \int_0^H \alpha\,dz}$$

where T_S is the surface temperature, $T(z)$ the air temperature at a height z, α the attenuation coefficient in the atmosphere, Θ', φ' the angular coordinates of the incident beam, Θ the viewing angle, φ the viewing azimuth, r $(\Theta_1,\varphi_1,\Theta',\varphi')$ the surface brightness coefficient, H the height at which the sensor is located, and z the vertical coordinate.

The analysis of expression (4.1) shows, first of all, that the multi-factor nature of the recorded $T_{B,\lambda}$ results in a limited accuracy of retrieval of a geophysical parameter. For example for the case of passive microwave sounding of the sea and land ice cover, the problem is further hindered by the fact that the ice emissivity Σ is determined not only by ice structure, but also by the vertical distribution of ice temperature and its concentration. The presence of liquid fraction (either salt-containing or fresh water), air inclusions and pollutants also changes substantially the dielectric parameters both of sea and fresh-water ice.

4.2 CALCULATION OF THE DIELECTRIC PERMEABILITY AND EMISSIVITY OF FRESH AND SEA WATER

Both water and ice are characterized by electronic, dipole-relaxation, and elastic-dipole polarizations (Debye, 1957), which leads to a substantial dependence of their emitting properties on the frequency of the external electric field. This special feature, on the one hand, introduces new limitations to the accuracy of the solution of inverse problems in satellite oceanography, and on the other hand, opens up new prospects for the spectral–polarization measurements and search for optimal conditions to retrieve various geo-physical parameters. In this connection, let us consider the spectral dependence of electrophysical parameters of sea and fresh-water ice, as well as water and snow.

The electronic inertialess polarization of water and ice as polar (dipole) dielectrics, is determined by shifting the cloud of atoms, molecules and ions under the influence of the external field and takes about 10^{-15} s to become stabilized. The dipole-relaxation polari-zation observed in dipole liquid and gaseous dielectrics, consists in increasingly regulated location of dipoles. The time for its stabilizing is about 10^{-10} s. The electric-dipole polarization is connected with elastic turning (deformations) of dipoles at small angles, when the dipoles cannot rotate freely.

The polarization of the polar liquid, P, according to the so-called new theory by Debye, is determined by the parameters of the medium and the field:

$$P(\omega) = \frac{\dot{\varepsilon}-1}{\dot{\varepsilon}+2}\frac{M}{\rho} = \frac{4\pi N}{3}\left[\alpha_0 + \frac{\mu^2}{3kT} + \frac{1}{1+i\omega\tau}\right] \tag{4.2}$$

where $\dot{\varepsilon}$ is the dielectric permeability, M is the molecular mass, ρ is the density of liquid, N is the Avogadro number, α_0 is the average polarization of molecules on the assump-tion that there is no interaction between the molecules, μ is the constant dipole moment, k is the Boltzmann constant, T is the absolute temperature, τ is the relaxation time, that is, the time needed for the dominating orientation of molecules, caused by the external field, to be replaced by a random distribution, when the field is removed, $\omega = 2\pi\nu$, where ν is the frequency of the external field.

The complex dielectric permeability of water in the microwave spectral region can be written as

$$\dot{\varepsilon} = \frac{\varepsilon_s - \varepsilon_0}{1+i(\lambda_s/\lambda)} + \varepsilon_0 \tag{4.3}$$

Here ε_r and ε_i are the real and imaginary parts of the dielectric permeability, written, respectively, as:

$$\varepsilon_r = n^2 - \chi^2 = \frac{\varepsilon_s - \varepsilon_0}{1 + \omega^2 \tau^2} + \varepsilon_0 = \frac{\varepsilon_s - \varepsilon_0}{1 + (\lambda_s/\lambda)} + \varepsilon_0 \tag{4}$$

$$\varepsilon_i = 2n\chi = \frac{(\varepsilon_s - \varepsilon_0)\omega\tau}{1 + \omega^2 \tau^2} = \frac{(\varepsilon_s - \varepsilon_0)\lambda_s/\lambda}{1 + (\lambda_s/\lambda)^2} \tag{4.5}$$

where $\lambda_s = 2\pi c\tau$ is the critical wavelength, ε_s is the statical dielectric permeability, ε_0 is the optical dielectric constant, the sum of atomic and electronic polarizations, and λ is the emission wavelength.

The correctness of the Debye formulas for fresh-water has been confirmed experimentally (Debye and Suck, 1936; Fröhlich, 1960). At the same time, the ideas about the dielectric properties of sea water as an electrolytic solution of the salts of Na, K, Mg, Ba, and other elements, have long been ambiguous. The reason for the degree of variability of the data reported by various experts is high values at radiofrequencies of electric conductivity of the water solutions of electrolytes, as well as an ambiguity of theoretical assessments of the effect of the salts of the elements enumerated above on the specific electric conductivity of solutions and the Debye relaxation absorption.

A detailed analysis of these effects and their dependences on wavelengths has been made by Rabinovich and Melentyev (1970). Without dwelling upon the description of the assumed modifications of calculation schemes, we shall only indicate that the laboratory experiments with the use of an original technique to measure the emissivity have verified the correctness of the Debye formulas for the water solutions of electrolytes with an equivalent concentration of NaCl salts to several hundredths of the normal solution on the assumption of techniques given by Kondratyev et al. (1992b) to take into account the dependences of ε_s, ε_0, and λ_s on the temperature of the solution and the concentration of salts.

The analysis of calculation data (Figs 4.1(a) and (b) showed that both ε_r and ε_i of sea and fresh-water depend strongly on temperature that the clearly expressed dependence on the salt content of waters increases with the growing wavelength, λ. The real part of the dielectric permeability of natural waters ε_r increases with increasing λ, for fresh-waters this increase manifesting itself more clearly. The absolute value of ε_r for salt waters in the 0.3–18.0 cm wavelength region is less than that of ε_r for fresh-waters at the same temperature. This difference manifests itself most at $\lambda > 2.0$ cm.

The variation in ε_i for fresh-waters is determined by the Debye absorption, whose maximum value at a water temperature of 0°C lies near $\lambda = 3.0$ cm. With the increasing temperature, this maximum shifts towards shorter wavelengths, the absolute value of ε_i being less in the whole wavelength interval under consideration. The appearance of salt in the water changes its electric properties in the region of the Debye absorption. A maximum of absorption, clearly expressed for each temperature, remains also for salt waters, but its absolute value decreases. With the wavelength increasing, the effect of ion conductivity of the solution shows, which breaks the former temperature-dependence of ε_i. The effect of ion conductivity in the interval of wavelengths longer than 1.5–3.0 cm (depending on the temperature of the solution) leads to an increase of the absolute value of the imaginary part of dielectric permeability with the increasing concentration of salts at any water temperature.

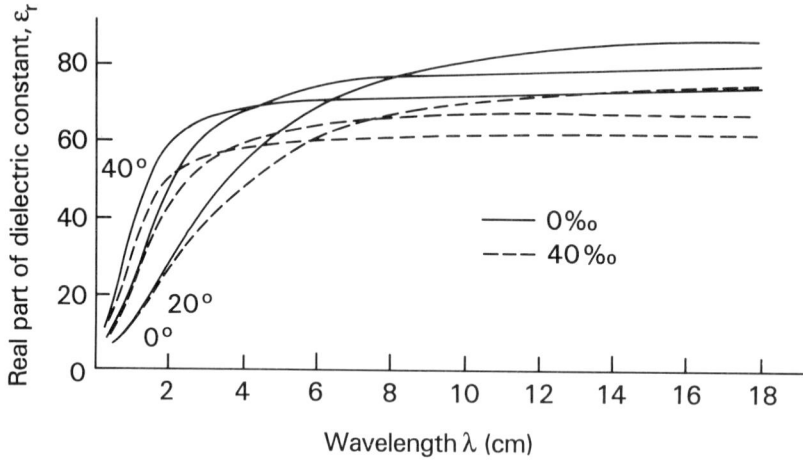

Fig. 4.1(a). Real part of dielectric constant of fresh and saline water as a function of wavelength, with temperature as the parameter (Rabinovich and Melentyev, 1970)

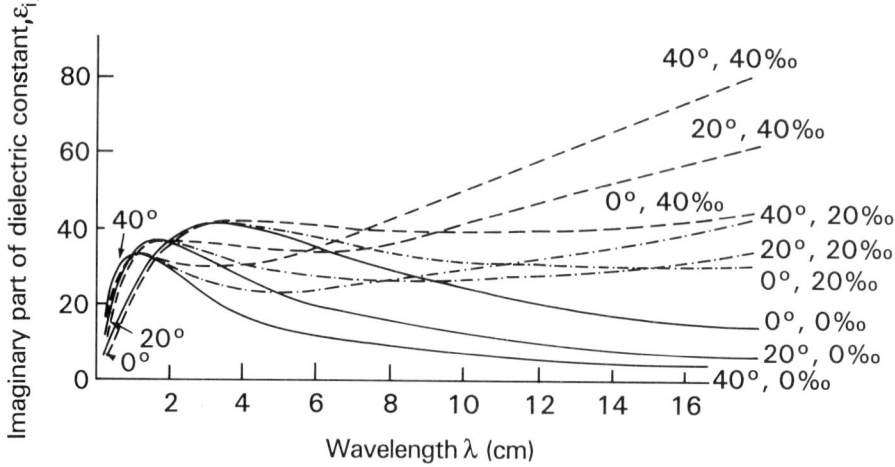

Fig. 4.1(b). Imaginary part of dielectric constant of fresh and saline water as a function of wavelength, with temperature as the parameter (Rabinovich and Melentyev, 1970)

The results of model calculations of dielectric parameters and emissivity of fresh and mineralized water surfaces have been tabulated by Rabinovich and Melentyev (1970). Note, that the data by these authors are used by most of the Russian experts when developing the technique of microwave remote sensing of water objects as the background ones.

This calculation scheme suggested in this publication has been applied to decimetre and metre wavelengths by Raiser et al. (1975). The results of these calculations show further increase of the dependence of ε_i on salinity with the growing wavelength, which can be used to substantiate the method of microwave sounding of the mineral content of natural waters.

4.3 TECHNIQUES FOR MEASURING SEA ICE PERMEABILITY UNDER CONTROLLED CONDITIONS

While for the negligibly elastic water the mechanism of the dipole-relaxation polarization is not observed in the wavelength region 10^9–10^{10} Hz, in the case of ice this mechanism is active in the region 10^2–10^4 Hz (Eder, 1947). It can be seen from Figs 4.2 that the ε_r of fresh-water ice changes drastically, starting from the acoustic frequencies. In the alternative field the changes in polarization lag behind the changes in the field, since the shiftings of changes in a solid dielectric cannot be instant. The difference in the phase of oscillations of the external field and the total vector of dipole moments of the ice crystalline grid expressed in radians is called the angle of the ice dielectric loss, δ. It turns out here that the tangent of the angle of loss, for which the formula

$$\mathrm{tg}\,\delta = \frac{(\varepsilon_s - \varepsilon_0)\omega\tau}{\varepsilon_s + \varepsilon_0\omega\tau} \tag{4.6}$$

holds, has a clearly expressed maximum shifting with air temperature increasing towards higher frequencies.

At the same time, it should be borne in mind that the polarization theory has some substantial limitations to its applicability, which is connected, first of all, with a

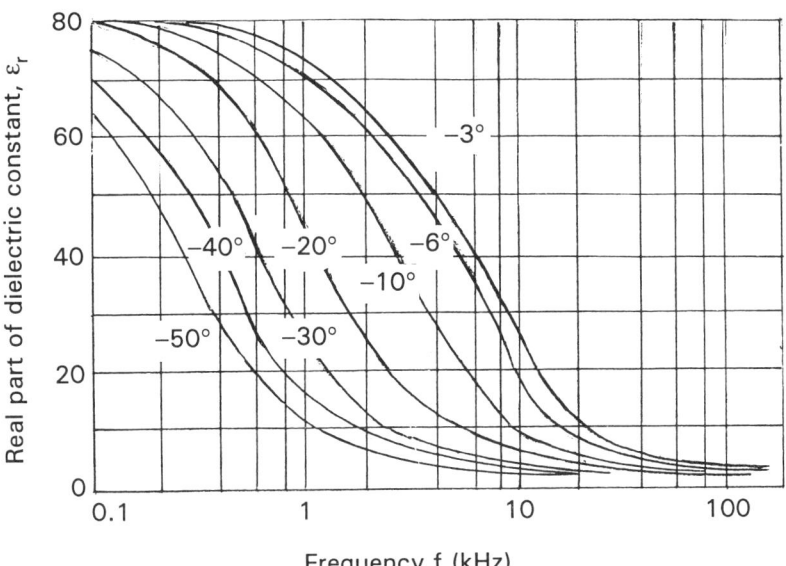

Fig. 4.2(a). Real part of dielectric constant of fresh ice as a function of frequency, with temperature as the parameter (Eder, 1947)

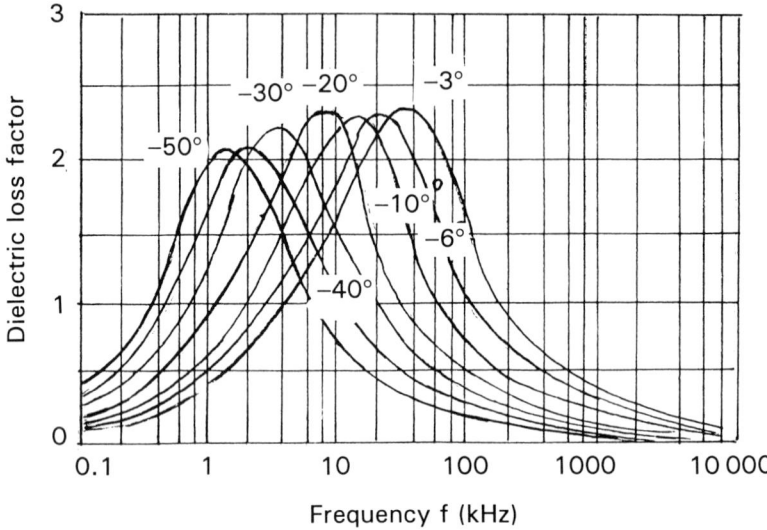

Fig. 4.2(b). Dielectric loss factor of fresh ice as a function of frequency, with temperature as the parameter (Eder, 1947)

simplified model representation of the behaviour of dipole molecules in the external field. Therefore from the viewpoint of the development of efficient techniques for the remote sensing of the parameters of water, ice, and snow, it is important to develop a technique for the observational studies of electric parameters of real natural structures and formations. The procedures to measure the electrophysical properties of dielectrics have been successfully developed (Brand, 1963; Hippel, 1959).

Let us briefly discuss the principal techniques of measuring ε_r and ε_i of ice, snow and soils for negative temperatures (Bogorodsky and Oganesian, 1987), since limitations inherent to any measurement procedure are quite essential in comparing the data of subsequent calculations and their comparison should be made carefully in the thematic interpretation of the remotely sensed information.

(1) The radar sounding technique based on receiving the wideband radar signal from the ice boundary layers. The antenna systems located either on the ice surface or at a small height over it enable one to study the reflectance of the boundaries, the average rate of radio pulses through the ice without breaking the ice condition. The limitation of the procedure is the uncertain location of the bottom-reflecting boundary.

(2) The double-line technique. Studies have been made of the natural ice cover using the measuring lines frozen-in at a certain distance from each other. The diameters of the lines and the distance between them are chosen from conditions of coordinated wave resistances of the measuring and feeding lines. The technique of short-circuit lengths enables one to obtain the rates of the vertical propagation of waves,

the effective value of ε_r, specific absorption, N, and the reflectance for the line length within the ice.

(3) The coaxial-line technique. It is used to study the electrophysical properties of the artificial ice samples. The space between two concentric conductors is filled with the salt solution of the concentration needed, and when the solution freezes, the line is sounded with video and radio pulses. The need for thorough monitoring of changes in the texture of the frozen ice limits the possibilities of the technique, since variations in the orientation of the cells of the salt solution with respect to vector E can change the fixed value of N by an order of magnitude.

(4) The microwave sounding technique. A specially prepared natural ice sample is located between the megaphone antennas or waveguides. During the measurements the length of the laboratory sample varies, the temperature and salinity varies, too. Unfortunately, despite the thorough control of the state of samples, a satisfactory agreement of the data of laboratory and field radar measurements cannot be obtained. It is connected with the phenomenon of near-electrode polarization and the edge-effects in the distribution of the field.

(5) The technique of video- and radio-pulse sounding. It uses the bell-shaped pulses of duration from units to several thousand nanoseconds, which makes it possible to substantially raise the accuracy of the estimated parameters.

Note that the use of the data of experimental studies of the ice and snow electrophysical properties needs not only a thorough account of differences in the density and composition of samples, as well as differences in the temperature and stress, but also an account of the technique for measuring the ice-cover constants.

4.4 FIELD MEASUREMENTS OF FRESH-WATER ICE PERMEABILITY

A detailed overview of the results of measurements for undisturbed ice (not preserved at higher temperatures for a long time) has been made by Evans (1965). The conclusion from this overview about the good agreement of the data of measurements of ε_r of the fresh-water ice is favourable for the development of techniques for its remote sensing in the microwave region. In the analysis of the fixed values of ε_r we shall take data by Hippel (1959) as a reference point: at frequencies 3.0 and 10.0 GHz the ε_r of the ice constitutes 3.20 and 3.17 (the temperature of the sample is –12°C). Unfortunately, the information about the frequency-dependence of the tangent of the dielectric loss angle given by various authors is widely scattered.

Detailed tables of the values of real (n) and imaginary (χ) parts of the coefficient of refraction of the fresh-water polycrystalline ice in the spectral interval from the UV to metre wavelengths have been published by Warren (1984). This author gives a detailed summary of such data for practically the whole microwave range: 1.0 mm–5.0 m. The ice temperature variability ranges from –1°C to –60°C.

As can be seen from calculations data, the ε_r depends weakly on temperature and varies from 3.16 to 3.20 as a function of wavelength and temperature, which agrees well with the published data for fresh-water ice. The tangent of the loss angle is, on the other

Table 4.1. Value of the constants in the formula for
approximate calculations of tg δ

λ, mm	A	B	C
1.0	0.004 26	0.115	0.006 90
10	0.002 60	0.174	0.001 40
20	0.002 20	0.191	0.000 84
40	0.001 40	0.185	0.000 85
90	0.000 30	0.097	0.000 24
120	0.000 16	0.093	0.000 20
180	0.000 03	0.023	0.000 06

hand, a characteristic that strongly varies, depending on these parameters. The scattering of tg δ values for millimetre wavelengths is particularly substantial.

For practical purposes, a simple formula has been recommended:

$$\mathrm{tg}\delta = A\, e^{-B|t_c|} + D \tag{4.7}$$

where A, B, D are the wavelength-dependent coefficients given in Table 4.1 and $|t_c|$ is the absolute value of ice temperature in degrees Celsius.

4.5 LANDAU–LIFSHITS FORMULA AND OTHER FORMULAS FOR MIXTURES

The ice and snow cover of seas and inland water basins usually has a heterogenic structure; therefore, to compare the data of field experiments with those of calculations, the emitting properties should be calculated for various mixtures.

In accordance with Netushil (1975), for a small concentration Θ_2 of one of the components for the real part of dielectric permeability, the formula by Landau–Lifshits has been recommended as it optimally describes electrophysical parameters of the real natural structures with negative air temperatures:

$$\varepsilon_{\mathrm{rmix}} = \varepsilon_{r1} + \Theta_2 \frac{3(\varepsilon_{r2} - \varepsilon_{r1})\varepsilon_{r1}}{\varepsilon_{r1} + 2\varepsilon_{r2}} \tag{4.8}$$

For the chaotically packed (statistical) mixtures of dielectrics, in which the grains of the components are located irregularly, the formula by Lichtenecker was used:

$$\lg \varepsilon_{\mathrm{rmix}} = y_1 \lg \varepsilon_{r1} + y_2 \lg \varepsilon_{r2} \tag{4.9}$$

where y_1, y_2 are the volume concentrations of the components.

For the dielectric particles of anisotropic shape, the formula by Weiner (1910) is recommended:

$$\frac{\varepsilon_{\text{rmix}} - 1}{\varepsilon_{\text{rmix}} + U} = p \frac{\varepsilon_{\text{rl}} - 1}{\varepsilon_{\text{rl}} + U} \tag{4.10}$$

where U is the parameter of the shape and p is part of the volume, occupied by monolytic ice (for the ice cover).

For spherical particles $U = 2$, for strongly stretched, parallel particles, $U = \infty$ was recommended by Basharinov et al. (1974) for the binary water–ice mixtures, $2 \leqslant U \leqslant 10$.

The formula by Weiner can also be used to calculate the imaginary part of dielectric permeability of mixtures; however, only its application to the real values of ε has been strictly justified.

The dielectric properties of sea ice are much determined by specific features of its size distribution, which depend on the conditions of its formation and age. Fresh sea ice several tens of centimetres thick has a large volume moisture content and an increased salinity. The 2-year and multi-year ice (several metres thick) is characterized by substantially less water content (salt solution) compared to fresh ice. The upper layers of 2-year and multi-year ice contain air bubbles and dry gaps. So, for example, the average ice porosity estimated from the air content constitutes about $50 \text{ cm}^3 \text{ kg}^{-1}$. With a strong porosity it can increase to $200–400 \text{ cm}^3 \text{ kg}^{-1}$.

The land ice formations (surface glaciers and permafrost zones) have multi-year history of their formation and are, respectively, from tens to hundreds of metres thick. The seasonal character of their transformations determines their stratified structures. There are vertical splittings, hollows, and cracks in the structures of glaciers, and permafrost is characterized by sub-ice water lenses (melted spots), both penetrating the whole permafrost layer and closed below (pseudo-melted spots). Note, that in accordance with Vtiurina (1970), the Russian classification of frozen soils is based on special cycles of their freezing–melting.

The short-lived frozen soils are characterized by the diurnal cycle of freezing–melting, at least, annually. The seasonally frozen soils have the seasonal cycle of freezing–melting, at least, once a multi-year period (assumed to last 30 years). The multi-year frozen soils are the aggradation phase of development of multi-year cryogenic soils, that is, soils characterized by only the multi-year freezing–melting cycle. The dielectric properties of such structures should be calculated from the formula for mixtures, since, as has been shown by England and Johnson (1975), the line of freezing divides the material into frozen and non-frozen constituents, different as to their electric properties.

The surface of land ice formations and frozen soils are, as a rule, snow-covered, which further complicates the simulation of their temperature, moisture content, salinity, porosity and density; but they can also be calculated from the formulas for multi-component mixtures.

4.6 STUDIES OF INTERRELATIONSHIPS BETWEEN ELECTROPHYSICAL AND PHYSICO-MECHANICAL PROPERTIES OF ARCTIC SEAS ICE COVER

The sea ice is a three-component system with varying properties. The seasonal variability of the state of ice formations is determined by varying thermodynamic conditions of the sea ice existence, as well as their freshening, resulting from the migration of salts through the thickness. The dielectric properties of sea ice, therefore, are characterized by strong spectral variations.

A number of researchers (e.g., Gorsky, 1957; Hoekstra, 1970) have shown that the emitting properties of sea ice depend strongly not only on the volume but also on the form and orientation of inclusions of the salt solution. They show that the most accept-able simulations of the behaviour of sea ice in the electric field can be achieved by considering the form of salt inclusions as ellipsoids of rotation with the ratio of axes taken to be 1:20. Real sizes of such ellipsoids are 0.5 to 10 mm and their long axis has an angle of inclination to the vertical varying within the range 35–45°.

A limited number of accepted petrographic descriptions of the indicators of the natural ice structures (shape, size, orientation, mutual position of crystals and mixtures) and their combination with the texture indicators make it possible to only qualitatively simulate the sea ice genesis, to show the peculiar climatic impact on ice formation and the character of dynamo-metamorphism. This circumstance can lead to an ambiguous correspondence of traditional oceanological classifications of ice structures to their quantitative microwave signatures. For these reasons, and because of difficulties of high-frequency measure-ments, the few publications containing information about the dielectric properties of sea ice, give widely scattered estimates.

Fig. 4.3, according to Bogorodsky and Oganesian (1987), shows the distributions of the temperature and salinity of fresh, one-year and multi-year ice of the Arctic Basin, typical of March. As can be seen from the data, fresh and first-year ice is characterized by the C-shape distribution of salinity. The distribution of salinity for multi-year ice is almost linear, over some areas, with some gradient. The temperature profiles are more smooth and are practical linearly dependent of ice thickness. The experts of the Arctic and Antarctic Research Institute (St. Petersburg) have obtained a large amount of obser-vation on the arctic ice-cover climatology, useful in the preparation of the data of the airborne and space-borne microwave sounding of the ice and snow cover.

Many researchers have studied the character of variations in the volume of liquid phase within the sea ice, especially in its boundary layers as most responsible for attenuation of electromagnetic energy. They suggested some parameterizations of this phenomenon. Fig. 4.4 illustrates the limits to variations in the liquid phase volume, characteristics of December to April in the arctic seas, and a parameterization of its averages by a polynomial to the third power (Bojarsky, 1983). Three values of the polynomial coefficients were calculated for each type of ice, corresponding to extreme and average approximated parameters.

The following expressions for the polynomial coefficients for fresh ice exemplify the data obtained:

$$V(H)_{min} = 0.501\,53H - 0.176\,95 \times 10^{-1}H^2 + 0.171\,86 \times 10^{-3}H^3$$

$$V(H)_{min} = 2.7079 + 0.623\,12H - 0.269\,39 \times 10^{-1}H^2 + 0.299\,27 \times 10^{-3}H^3$$

$$V(H)_{min} = 3.5196 + 1.4325H - 0.534\,53 \times 10^{-1}H^2 + 0.552\,43 \times 10^{-3}H^3$$

$$(4.11)$$

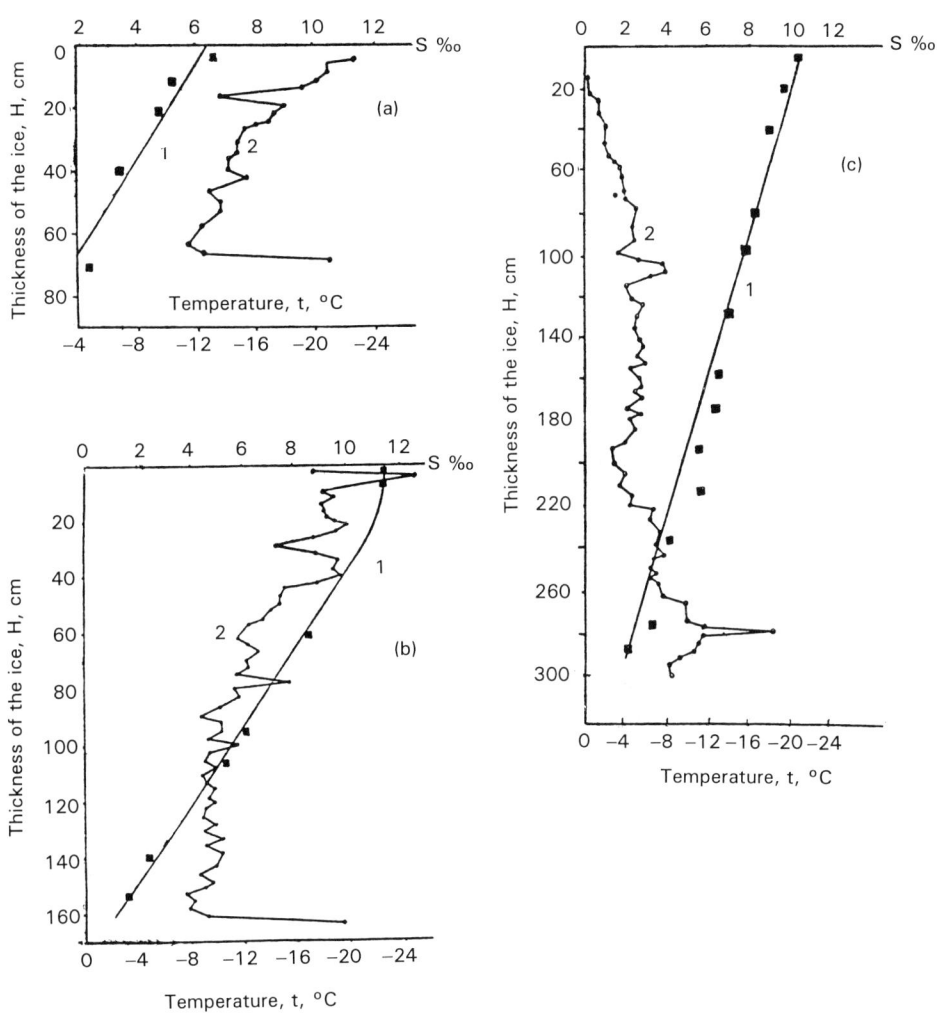

Fig. 4.3. Vertical distribution of the temperature (1) and salinity (2) of sea ice (March): (a) young
ice; (b) first-year ice; (c) multi-year ice (Bogorodsky and Oganesian, 1987)

Similar dependences are available for the first-year and multi-year ice.

In connection with the especially important problem of the remote sensing of sea ice
of different ages, the problem of an adequate parameterization of the electric characteristics of sea ice of different thickness and age remains particularly urgent. Let us consider
the results by Darovskikh (1984) who analysed the experimental data on the electric
parameters of sea ice of different thicknesses and showed that ice with a large liquid
phase volume ($V > 10\%$) can be considered as binary mixtures containing the salt solution and ice crystals. In this case the refraction index, n, and specific attenuation N of the
mixture with electric parameters of salt solution n_{ss} and N_{ss} can be described with the
following relationships:

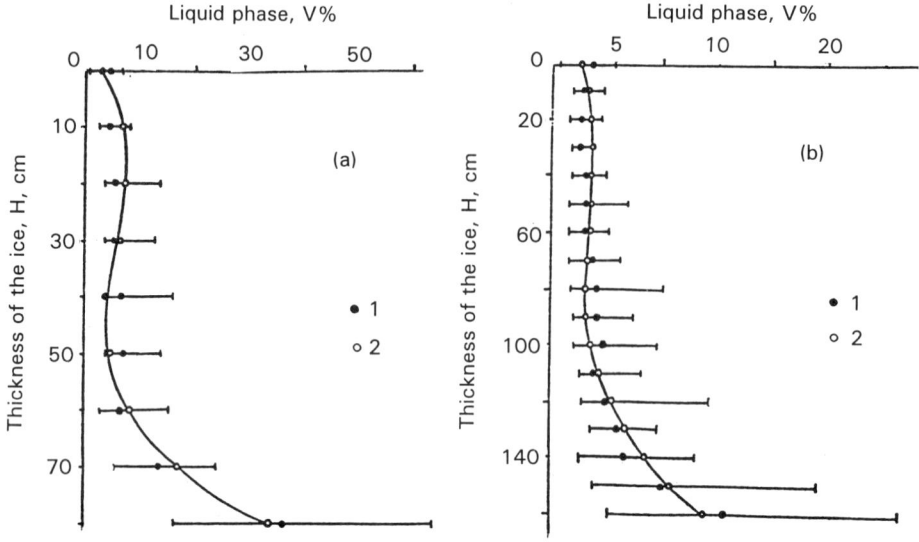

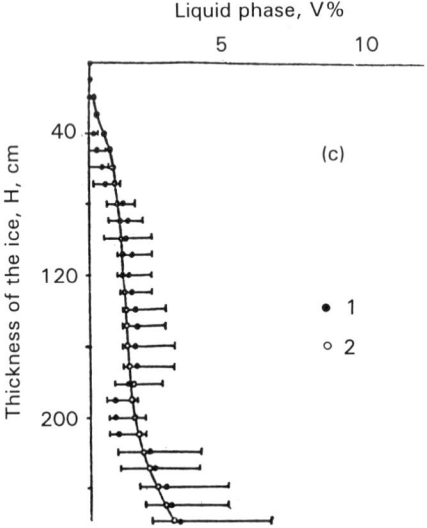

Fig. 4.4. Vertical distribution of the volume of the liquid phase for: (a) young ice; (b) first-year ice; (c) multi-year ice; 1, mean value of V and limits of the variation for December–April; 2, polynomial approximation (Bogorodsky and Oganesian, 1987)

$$n = 1.78 + (n_{ss} - 1.78)V$$
$$N = N_{ss}$$
(4.12)

The values of n_{ss} and N_{ss} determined from the temperature and salinity of salt solution are calculated from relationships suggested by Stogryn (1984).

To calculate the electric characteristics of thicker one-year ice with the liquid phase volume $V < 10\%$, Morozov and Khokhlov (1973) suggested linear dependences of n and N on the content of liquid phase:

$$n = 1.78 + aV$$
$$N = bV$$
(4.13)

For the wavelength region $0.72 \leqslant \lambda \leqslant 20.0$ cm, the experimental studies give the following approximative dependences of the coefficients a and b on λ:

$$a = 1.132 \ln(5.326\lambda - 1.006\,52)$$
$$b = 60.51\lambda - 0.8162$$
(4.14)

An analysis of estimates of n and N in the upper layer of the one-year sea ice suggests that the ice of different thicknesses for each H_A has an individual interval of variations in the electric parameters, which forms the physical basis for microwave sensing of the age characteristics of sea ice. Besides, according to Bogorodsky et al. (1983), a strong attenuation of electromagnetic waves makes it possible to neglect the scattering of emission on various internal inhomogeneities of the first-year ice, and results in the thickness of the layer, in which microwave emission practically totally forms, turning out to be less than the ice-cover thickness. Therefore, the first-year sea ice in the microwave region can be presented as an emitting vertically inhomogeneous, non-isothermic, snow-covered semi-space. The thermal emission flux for such a system is the sum of two fluxes: the ice emission through the snow layer and the snow-cover emission. It has been shown by the same authors that in the winter-to-spring period the microwave emission of the snow cover on fresh arctic ice at negative air temperatures can be neglected.

Figs 4.5(a)–(e) shows the data on the dielectric parameters of the sea ice cover of different structures in the joint Soviet–American "Bering Sea Experiment—BESEX" (Ramseier et al., 1975; Kondratyev et al., 1973a). The calculated dependences of the dielectric parameters of sea ice have been obtained for various sounding conditions (variations in the temperature and salt content of ice) and different parameterizations of the shape of particles (sphere, stretched particles). Of interest is a good coincidence of the data mentioned and the experimental data by Morozov and Khokhlov (1973) obtained with the use of an original technique for the determination of the dielectric constants of ice samples of the near-edge zone of the Bering Sea on board the ship "Priboy" (within the programme of the "Bering Sea Experiment"). The data given in Fig. 4.5 were used to interpret the aircraft microwave information. The calculations of the ice emissivity made on the basis of those data will be given below.

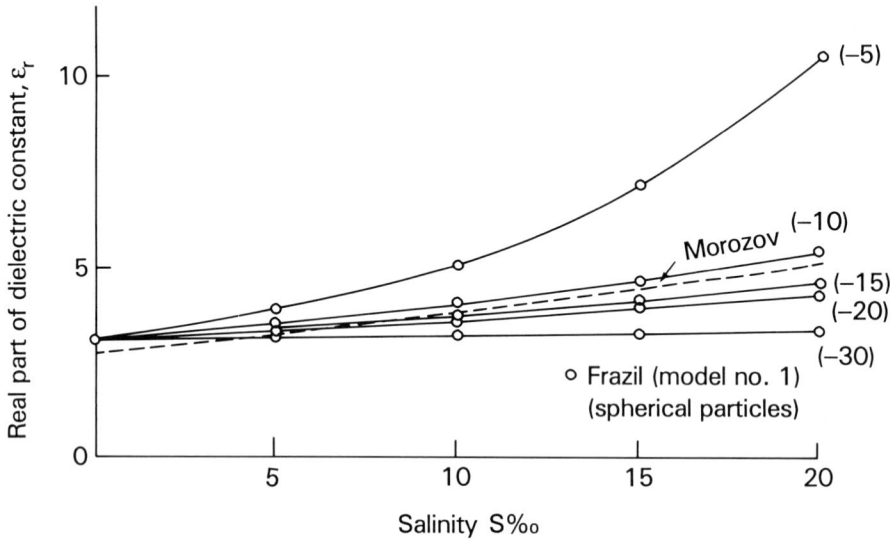

Fig. 4.5(a). Real part of dielectric constant of sea ice (Ramseier et al., 1975)

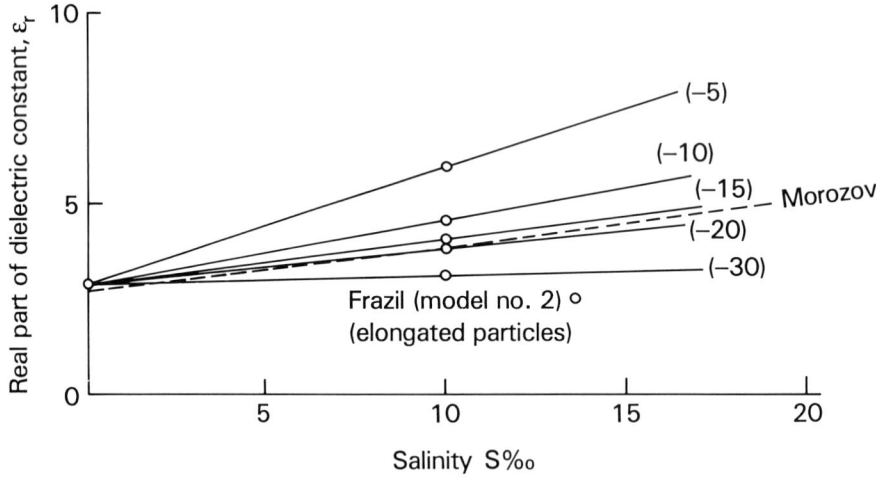

Fig. 4.5(b). Real part of dielectric constant of sea ice (Ramseier et al., 1975)

4.7 FINE STRUCTURE OF ELECTRICAL PROPERTIES OF SNOW-ICE COVER AND THEIR IMPACT ON ELECTROPHYSICAL PARAMETERS OF SEA ICE

The dry snow cover as as structure always present in the thematic interpretation of microwave information, as shown in numerous studies, can be considered as a mixture of ice and air. Since the real part of the ice dielectric permeability within the microwave

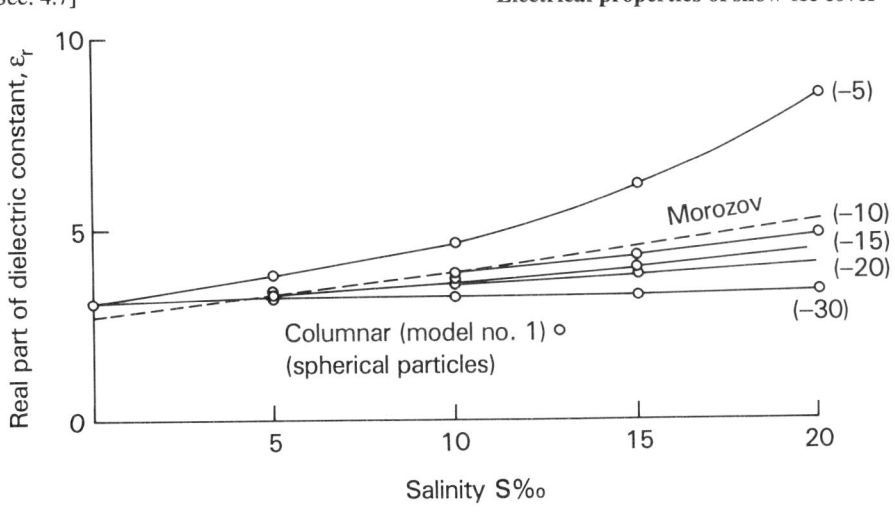

Fig. 4.5(c). Real part of dielectric constant of sea ice (Ramseier et al., 1975)

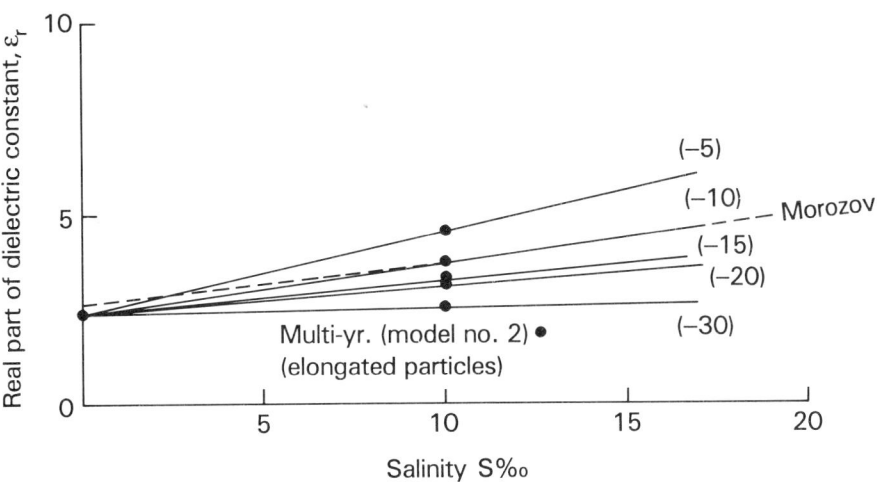

Fig. 4.5(d). Real part of dielectric constant of sea ice (Ramseier et al., 1975)

region is practically independent of temperature and wavelength, the parameter ε_r of dry snow at different λ is also constant, when the air temperature changes. First, this circumstance has been discovered by Cumming (1952), who shows that ε_r for dry snow is only a function of its concentration. It can be calculated with the use of the formula by Polder–Van Santon (1946):

$$\frac{\dot{\varepsilon}_{rds} - 1}{3\dot{\varepsilon}_{rds}} = \frac{V_{ice}(\dot{\varepsilon}_{rice} - 1)}{(\dot{\varepsilon}_{rice} + 2\dot{\varepsilon}_{rds})} \tag{4.15}$$

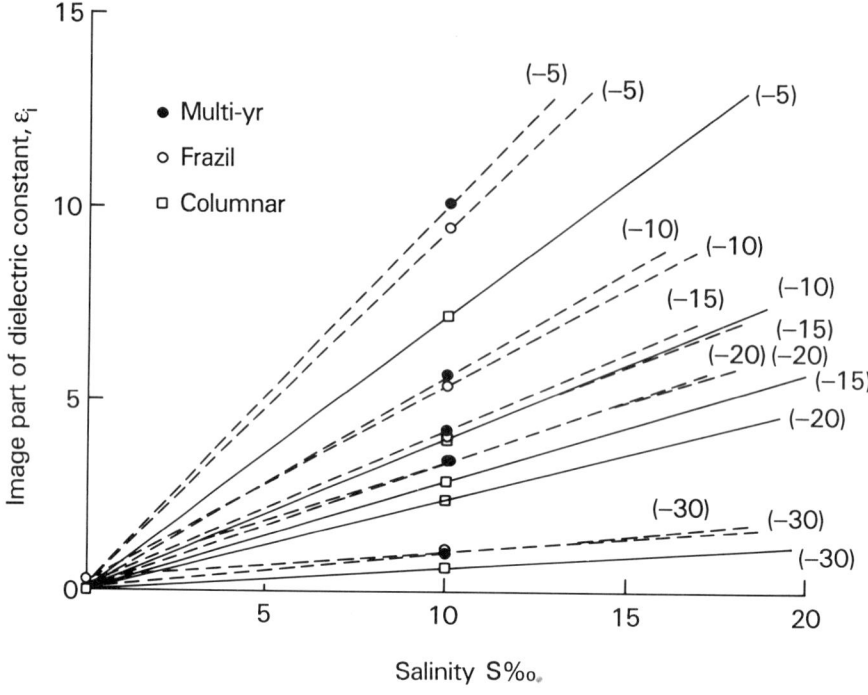

Fig. 4.5(e). Imaginary part of the dielectric constant of sea ice (Ramseier et al., 1975)

where $V_{ice} = \rho_{ds}/\rho_{ice}$ is the part of the total volume of dry snow covered with ice fraction and $\rho_{ice} = 0.916$ g cm^{-3} is the ice concentration.

The polarization effects taking place in case of snow cover sounding due to different geometry of ice particles fastened together and their orientation in the external field, ϖ, can be taken into account with the use of the formula by Weiner for the complex dielectric permeability $\dot{\varepsilon}_{mix}$ of two media:

$$\frac{\dot{\varepsilon}_{mix} - 1}{\dot{\varepsilon}_{mix} + u} = p\left(\frac{\dot{\varepsilon}_1 - 1}{\dot{\varepsilon}_1 + u}\right) + (1 - p)\frac{\dot{\varepsilon}_2 - 1}{\dot{\varepsilon}_2 + u} \tag{4.16}$$

where $\dot{\varepsilon}_1$ and $\dot{\varepsilon}_2$ are the complex dielectric permeabilities of the first and second media, p is part of the total volume occupied by medium 1, and u is the shape parameter.

The stratified snow structure, in which the electric field is normal to the layers, has $u \rightarrow 0$, and the structure, in which the electric field is parallel to the layers, has $u \rightarrow \infty$. In the case of small losses (tg$^2\delta \ll 1$), from expression (4.16), for the real part of dielectric permeability we have formula (4.10), and for the imaginary part we have

$$\varepsilon_{i,sn} = \frac{\varepsilon_{i,ice}(\varepsilon_{r,sn} - 1)^2}{p(\varepsilon_{r,ice} - 1)^2} \tag{4.17}$$

It is assumed that the measured values of the input parameters in which the parameter u has been taken into account, are used in formulas (4.10) and (4.17).

Detailed experimental and theoretical studies of the variability of dielectric properties of dry and humid snow have been made by Bogorodsky et al. (1971). Variations in ε_r and ε_i of snow have been studied in the region of variability of the determining parameters: water equivalent 0–12.3% of the volume of ice sample; snow density varied from 0.09 to 0.42 g cm^{-3}; temperature from 0°C to –15C; size of crystals from 0.5 to 1.5 mm. The measured data show that the dielectric properties of humid snow are closely connected with the dielectric properties of water—their variability with frequency. For dry snow the contribution of the volume scattering takes place at a frequency of 37.0 GHz, mostly used in the experiment. The use of some empirical and theoretical parameterizations of the mixture models shows that both the Debye semi-empirical model and the mixture formula by Polder–Van Santon provides a good agreement of the measured data for the real part of the dielectric permeability of wet snow only when the shape of water inclusions turns out to be asymmetric and depends on the snow water content.

Fig. 4.6 summarizes the measurement data on the variability of the real part $\varepsilon_{r,ws}$ of wet snow, depending on snow density for a broad interval of frequencies—the parameterization of the data with the model by Tinga et al. (1973). As can be seen from Fig. 4.6, a good agreement is observed between the measurement data and both models, but for the second model it is closer.

Figs 4.7 and 4.8 illustrate the dielectric parameters ε_r and ε_i for wet snow as a function of frequency with varying content of liquid phase, calculated in accordance with the modified model by Debye (Hallikainen et al., 1986). The dependences are given for the wet snow density $\rho_{ws} = 0.24$ g cm^{-3}—an average value of this parameter. It is seen from these data that $\varepsilon_{i,ws}$ follows the behaviour of liquid water at 0°C demonstrated above.

Of interest are the results of studies on the effect of the shape of water inclusions in snow: new formations of water in snow are needle-shaped. Its transformation takes place at $m_v = 3\%$, when the inclusions start taking a disk shape. These phenomena of transformations of the shape of water inclusions must show themselves with the polarization selection of microwave signal.

Now we shall discuss one more important aspect of the problem of the remote sensing of multilayer structures of the type "water–(frozen soil)–ice–snow": the problem of the formation of the temperature contrasts of such structures. Based on their problem-oriented field experiments and calculated data, the temperature contrasts have been studied in detail, typical of the surfaces of polar region, both for stationary hydrometeorological conditions and for the case of the effect of the variability of radiative regime and sensible snow-ice surface heat exchange with the atmosphere. The interaction of the principal factors, which form the microwave emission of the snow-ice cover of the arctic seas in the autumn–winter period, with apparent surface temperature contrasts has been studied based on the model of a 1-D stationary heat conductivity of the water–ice–snow–air system. The equations of heat balance and the equality of the temperatures of adjacent media were taken as boundary conditions.

The calculations assumed the absence of ice growth at the water–ice interface. The horizontal heat loss in the ice and snow and the effect of internal heat sources connected with phase transformations in the sounded layer were neglected. The solution of the

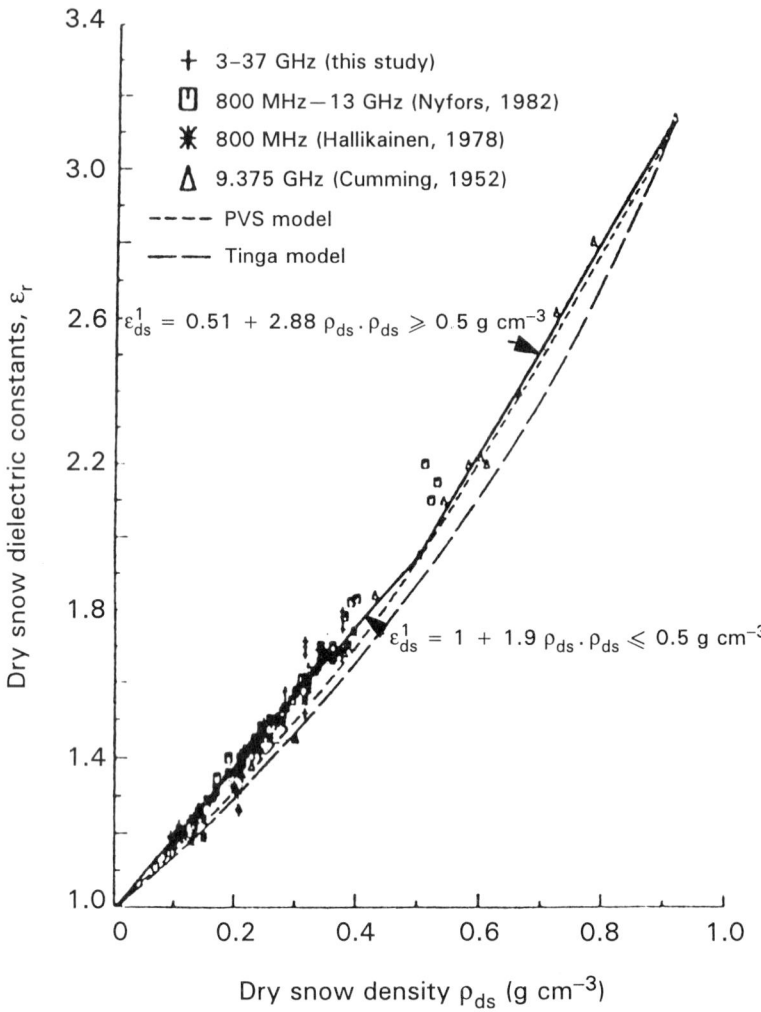

Fig. 4.6. Variation of dry snow dielectric constants with dry snow density (Hallikainen et al., 1986)

equation of heat conductivity gave an expression to calculate the distribution of the surface temperature of the ice cover of different thicknesses and snow cover in a given air temperature interval under conditions of changing hydrometeorological conditions:

$$T_s = \frac{H'}{1 + \delta H} \left(\frac{Q}{H'} + R + \delta T + \beta \psi \, e^{\alpha T_{air}} \right) \qquad (4.18)$$

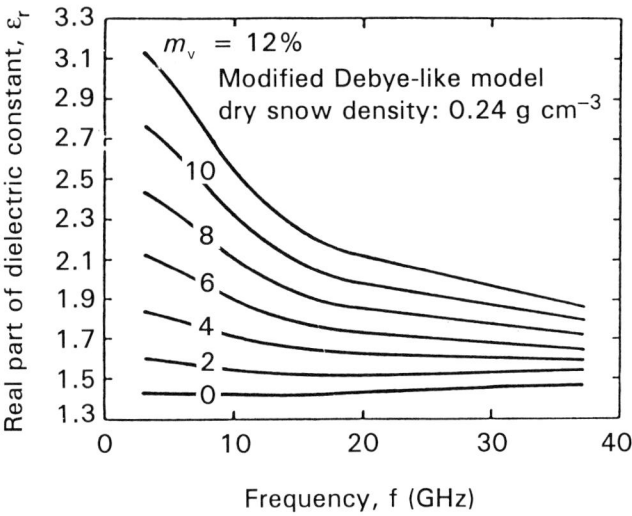

Fig. 4.7. Dielectric constant of wet snow according to the modified Debye-like model plotted as a
function of frequency, with liquid-water content as a parameter (Hallikainen et al., 1986)

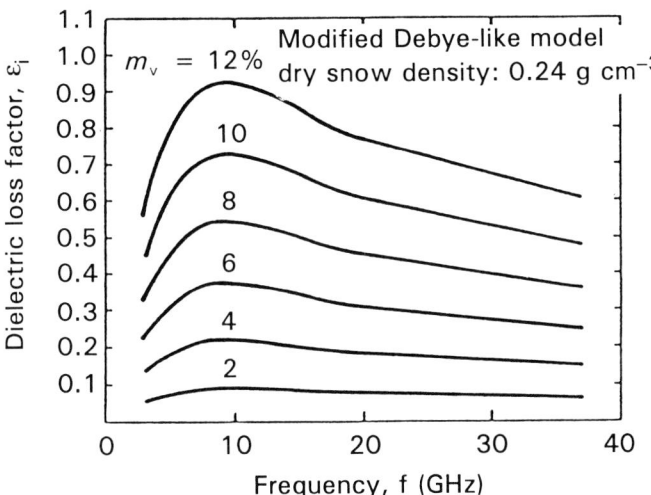

Fig. 4.8. Loss factor of wet snow according to the modified Debye-like model plotted as a
function of frequency, with liquid-water content as a parameter (Hallikainen et al., 1986)

where

$$H' = \frac{H}{\lambda_{ice}} + \frac{h}{\lambda_{sn}}$$

is the thickness of the ice-snow two-layer medium,

$$\delta = \beta(1+\psi); \quad \beta = \frac{C_{air}\beta_{air}k'}{\ln k/\eta}; \quad k = \eta + k'z$$

is the coefficient of sensible heat exchange,

$$\psi = 0.622\frac{E_0}{P}\frac{L}{C_{air}}(r-1)$$

H, λ_{ice} are the thickness and the coefficient of heat conductivity of ice, Q is the temperature of water freezing, R is the radiation budget; T_{air} is air temperature, C_{air} is the air heat capacity, ρ_{air} is the air density, η is the coefficient of sensible heat exchange for $z = 0$, L is the latent heat (snow condensation), q is the specific air humidity, E_0 is the water vapour pressure at 0°C, P is atmospheric pressure, $\alpha = 0.086$ deg^{-1}, r is the relative air humidity, and z is the vertical coordinate.

Calculations of the surface temperature of the flowing snow-ice cover for the ice thicknesses between 0.1 and 4.0 m and snow thicknesses between 0 and 60 cm, with temperature variations from 0 to –40°C, have shown considerable temperature contrasts of thin ice of different thicknesses and their gradual smoothing for ice more than 1.0 m thick (Fig. 4.9). The temperature contrasts determined by differences in the ice thickness are governed by air temperatures; with air temperatures decreasing the contrast grows. The presence of snow on the ice surface reduces the dependence of the snow-covered ice temperature on its thickness. With the snow layer thicker than 20 cm the temperature contrasts are practically absent.

From the estimates of the effect of the thermal–physical characteristics of snow on the thermal (and microwave, respectively) contrast of the snow-ice surface, it follows that the screening effect of snow depends much on its density (heat conductivity). Fig. 4.10 shows the dependence of the temperature of the snow-covered ice surface on its thickness, with different thicknesses of three types of snow cover: freshly fallen, condensed fine-grained, and firnized, with the coefficients of heat conductivity $(4.9, 7.1, 9.3) \times 10^{-4}$ cal cm^{-1} s^{-1} deg^{-1}, respectively. The calculations confirmed by the data of the infrared survey show that the screening effect of snow cover is at a maximum when the snow is fresh.

The heat conductivity of the sea ice itself varies within smaller limits than that of snow. Calculations have shown that the difference in the coefficient of the heat conductivity of the snow-covered sea ice of different age practically does not tell on the thermal contrast of such structures. In the case of snow absence the difference in the temperatures of the adjacent areas of sea ice, even with extreme coefficients of heat conductivity, constitutes not more than a tenth of one degree (slightly increasing with the temperature rising). This circumstance substantially favours the technique of the microwave sensing of the thermal regime of the ocean–atmosphere boundary layer in the polar regions, substantially simplifies the development of the technique for thematic interpretation of the parameters of the state of multi-year arctic structures.

For calculations of the snow-cover emitting properties, in the analysis of the data of aerospace remote sounding, the data on the snow cover climatology in polar regions are very important. Fig. 4.11 shows the histograms of the vertical distribution of snow cover

characteristics of ice formations of different age, based on processing the regular observational data at polar hydrometeorological stations and observatories in Russia. The
observational data for the period of 8–10 years were processed using the WMO standard
techniques.

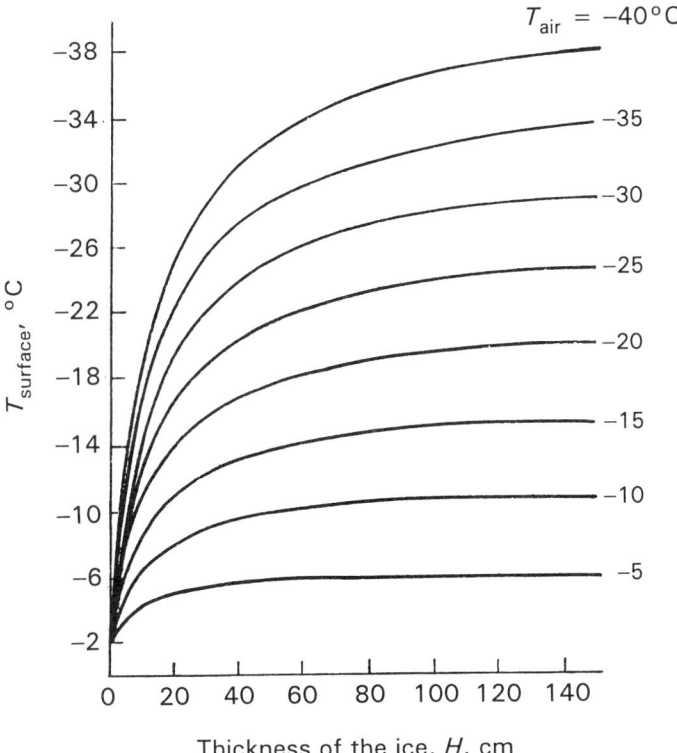

Fig. 4.9. Distribution of the ice surface temperature (without snow cover) (Bogorodsky and
Martynova, 1978)

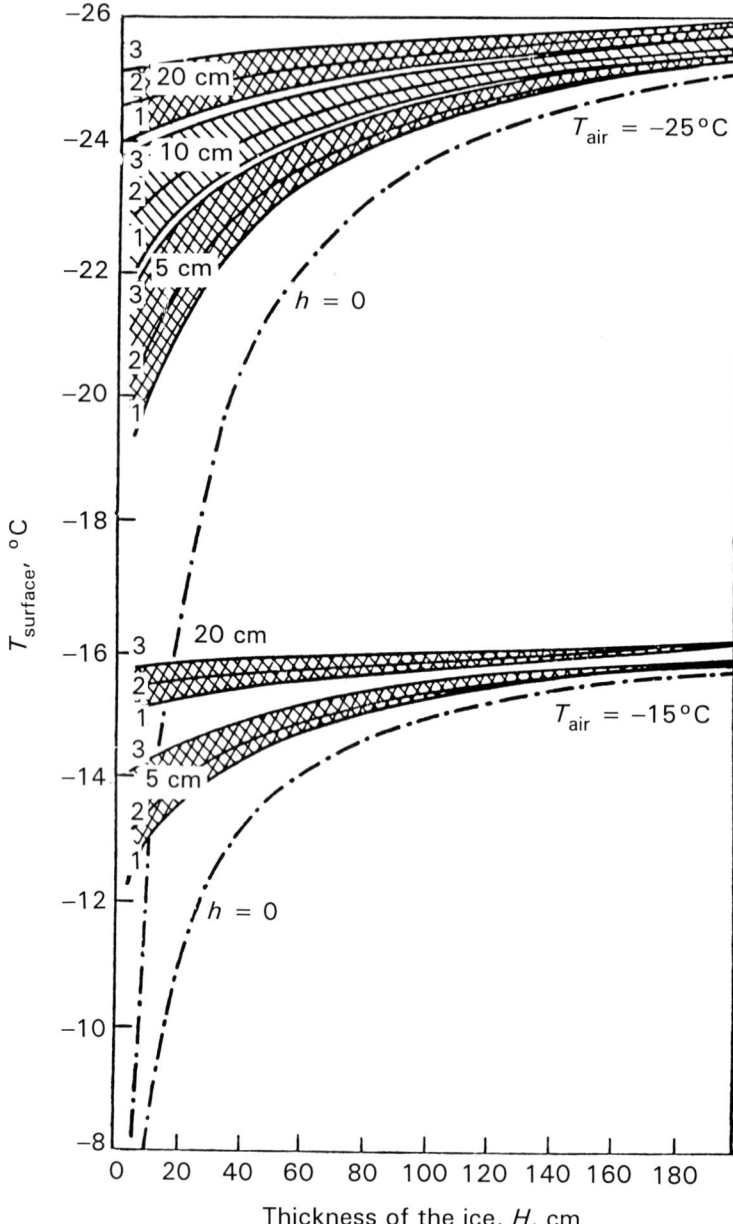

Fig. 4.10. Distribution of the temperature of snow on the sea ice (for the different thickness and density of snow cover) (Bogorodsky and Martynova, 1978)

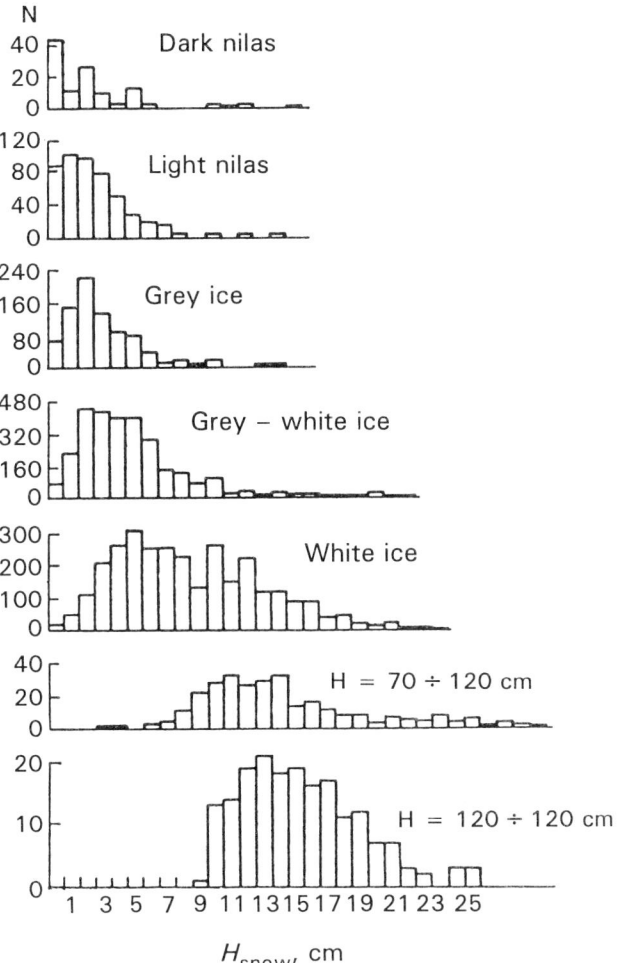

Fig. 4.11. Distribution of the thickness of snow on the ice different age (Bogorodsky and Martynova, 1978)

5

Numerical modelling of the spectral-polarization properties of fresh-water and sea ice, snow cover and frozen soil

5.1 PLANE-PARALLEL MODEL OF MICROWAVE EMISSION FOR NATURAL FORMATIONS AND STRUCTURES (PURE ABSORPTION APPROXIMATION)

A flat layer of a certain thickness "1" over the homogeneous semi-space is the simplest theoretical model of the surface for the remote sensing applications in satellite meteorology and oceanography. Despite an apparent extreme of this idealization, this model makes it possible to assess some phenomena, for example, the interface effects manifesting themselves through emission of such inhomogeneous structures as ice on the water surface, oil spills, foam and wake traces of ships, snow on land surfaces, etc. (Tuchkov, 1968; Kondratyev and Melentyev, 1994).

A sharp interface between the layers means physically that variations in the dielectric parameters take place over some length much less than the characteristic emission wavelength in a given medium. If the layer of ε variations is comparable with λ in thickness or exceeds it, then such a structure corresponds to a model of stratified structure with transition layers.

As has been shown by Basharinov et al. (1974), in the regular stratified structures the reflections from interfaces lead to clearly expressed resonance effects, especially with small values of the tangent of loss angle in the medium. At the same time, for most of real covers, the thickness of the layer is a random parameter which leads to the blurred oscillations, to correlations of inhomogeneities. The properties of the media with random inhomogeneities can, therefore, be considered using the models of a random stratified structure or a structure with random 3-D inclusions.

For the case of a flat homogeneous layer of a thickness "1", located at the boundary of a homogeneous medium, the dielectric properties of which differ from the properties of the layer, the following expression holds for the emissivityΣ:

$$\Sigma = \frac{\left[1 - R_{2,\lambda}\exp(-2\tau_\lambda)\right]\left(1 - R_{1,\lambda}\right)}{1 + R_{1,\lambda}R_{2,\lambda}\exp(-2\tau_\lambda) + \sqrt{R_{1,\lambda}R_{2,\lambda}}\,e^{-\tau_\lambda}\cos[2kl\cos v + \varphi_1 + \varphi_2]} \tag{5.1}$$

where $R_{1,\lambda}$, $R_{2,\lambda}$ are reflectances for the interfaces "air–layer" and "layer–homogeneous medium", respectively, $\tau_\lambda = \gamma_\lambda l/\cos v$, γ_λ is the coefficient of absorption in the layer, the angles v_0 and v are related through the Snellius law; $k = (2\pi n_\lambda)/\lambda$, n_λ is the index of refraction in the layer, the angles φ_1 and φ_2 are variations in the phase of the wave reflection from the interface "air–layer" and "layer–homogeneous medium", which depend on polarization and angle of incidence, v. It follows from (5.1) that the dependence of Σ on the parameter kl has an oscillating character, as well as the fact that the absorption in the layer smooths the oscillations.

Formula (5.1) was obtained for monochromatic emission. For the case of microwave sensor, with a broad band of transmission $(\lambda, \lambda + \Delta\lambda)$:

$$\Sigma = \frac{1}{\Delta\lambda}\int_\lambda^{\lambda+\Delta\lambda}\Sigma_\lambda\,d\lambda \tag{5.2}$$

With a broad-band sensor, when $\Delta\lambda/\lambda \gg kl\cos v$ and when oscillations get blurred even with a small absorption in the layer, the emissivity of a two-layer medium is determined as:

$$\Sigma = \frac{\left[1 - R_{2,\lambda}\exp(-2\tau_\lambda)\right]\left(1 - R_{1,\lambda}\right)}{1 - R_{1,\lambda}R_{2,\lambda}\exp(-2\tau_\lambda)} \tag{5.3}$$

In the case when the underlying layer has a considerable specular reflecting constituent (like, for example, the water in the microwave range), an appearance of the upper absorbing layer leads to the increasing blackness of the two-layer system. An increase of the absorption of a new-formed layer due to the increase of its length or smooth variations in its own properties, leads to a monotonic increase of Σ.

In the non-isothermic cases, the special shape of the spectrum of a fixed T_B turns out to be dependent of the sign of the temperature gradient of the upper layer. With a poor conductivity of the upper layer the T_B of the two-layer system can have either increasing or decreasing branches of the variability of the sign of the temperature contrast.

For most of the natural formations and structures (Basharinov et al., 1974), there is no sharp interface between the layers, and a satisfactory simulation of the character of their microwave emission is reached with the use of the models of flat structures with the transition layers, in which the dielectric permeability varies smoothly at one or several wavelengths. The transitional layers with smoothly varying parameters describe the decrease of reflectance at their interfaces and smoothing of the interference effects observed in natural conditions.

Bogorodsky et al. (1981) have shown that with the thickness of the transitional layer exceeding half a wavelength, there is no reflection at all. In these conditions the radiative properties of the model turn out to be close to the radiative properties of semi-space with the parameters of the upper layer. The roughness of the interface causes the radiation scattering and leads to the spatial variations of the efficient thickness of the layer, which is followed by smoothing the oscillating structure of the emission field. So, for example, within the faceted model of the sea roughness (Martsinkevich and Melentyev, 1972), the change in the facet inclination causes changes in the free path length and variations in phase shiftings. As a result, the interference structure is averaged over the set of phase shifts. In conditions of the observational experiment the spatial averaging with the directional aerial system is added.

The microwave emission of natural media containing random inhomogeneities of density and composition can have special features connected with the emission scattering on these inhomogeneities. Most solid dielectric structures are stochastically inhomo-geneous and to calculate their emissivity, statistical methods should be applied. According to the theory of general electrodynamics of thermal fluctuations (Rytov, 1953), the solution of the problem of emission of a medium with random inhomogeneities consist in calculations of active losses of the auxiliary wave. Calculations of such losses require the solution of the corresponding diffraction problem. In the isothermal medium it is sufficient to calculate the share of power of the auxiliary wave scattered by the medium to the environment. However, in this case, also, it is necessary to solve the problem of diffraction and to calculate the field outside the emitting structure. There is no solution, however, for such a problem in general, and therefore only approximate solutions are possible.

For the medium with random inhomogeneities of the refraction index the scattering on these inhomogeneities can be considered equivalent to the change of the coefficient of refraction from the medium–air interface. Then, according to Basharinov et al. (1974), the T_B of this structure can be written as:

$$T_B = T_0 \left[1 - (R_\lambda + \gamma_\lambda) \right] \tag{5.4}$$

where T_0 is the medium temperature; R_λ the coefficient of reflection at the medium–air interface; γ_λ the coefficient that describes the volume scattering and refers to the share of the power of the auxiliary wave mentioned above, which is scattered by volume inhomogeneities across the interface towards the environment and gets to the aerial of the microwave radiometric receiver.

The procedure of calculation of γ_λ is the simplest when the solution of the diffraction problem can be confined to single scattering with the use of the perturbation technique. In this case, according to Tatarsky (1967), with the medium–air interface assumed to be flat, one can study a number of specific cases of the state of natural formations and structures, using a sufficiently general expression for γ_λ.

To develop the algorithms to calculate the emissivity and the T_B of the multilayer systems of the type "water-ice-snow–air", and "water–ice–air" from the data on real distributions of the vertically inhomogeneous structures, a number of authors (Melentyev and Alexandrov, 1989; Kondratyev et al., 1989; Bogorodsky et al., 1977; Rabinovich et al., 1975; Gurvich et al., 1973) used a model of a smoothly inhomogeneous dielectric structure suggested by Stogryn. The form of the dependence of polarized emission on the vertical distribution of dielectric permeability and thermodynamic temperature suggested by Shulgina (Kondratyev et al., 1973b) was used in this case. It was shown that for the structure with the lathe value of attenuation, the T_B of the medium, which is proportional to the signal measured with the radiometer, can be calculated at the product of the effective temperature of the medium (T_{eff}) by the coefficient of emission (transmission) of the ice–air (snow–air) interface, q.

The calculations of the sea ice emissivity were made using the formulas:

$$\Sigma = \frac{\left[1 - R_{21}^2(\Theta)\right]\left[1 - R_{32}^2(\Theta)\,e^{-2\beta} - 4R_{32}(\Theta)R_{21}(\Theta)\,e^{-\beta}\sin(\varphi_{32}+\alpha)\sin\varphi_{21}\right]}{1 + R_{32}^2(\Theta)R_{21}^2(\Theta)\,e^{-2\beta} + 2R_{32}(\Theta)R_{21}(\Theta)\,e^{-\beta}\cos(\varphi_{32}+\varphi_{21}+\alpha)}$$

(5.5)

where R_{32} and R_{21} are the Fresnel coefficients of reflection at the interfaces "water–ice" and "ice–air", respectively, h is the ice layer thickness, and Θ is the viewing angle.

$$\beta = \frac{4\pi\chi_2}{\lambda}; \qquad \alpha = \frac{4\pi n_2}{\lambda};$$

(5.6)

$$\text{tg}\,\varphi_{j,k} = \frac{2(\delta_j\gamma_k - \delta_k\gamma_j)}{\gamma_j^2 - \gamma_k^2 - \delta_j^2 - \delta_k^2}; \qquad j,k = 1,2,3$$

(5.7)

For the horizontal polarization:

$$\delta_k = \frac{\tilde{\chi}_k}{\tilde{n}_k^2 + \tilde{\chi}_k^2} \qquad \gamma_k = \frac{\tilde{n}_k}{\tilde{n}_k^2 + \tilde{\chi}_k^2} \qquad k = 1,2,3$$

(5.9)

where ε_{rk} and ε_{ik} are the real and imaginary parts of the dielectric permeability of the respective media; λ is the emission wavelength.

The refraction index $\tilde{n}_k$ and the absorption index $\tilde{\chi}_k$ for $k = 1,2,3$ are

$$\tilde{n}_k = \sqrt{\frac{\varepsilon_{rk} - \sin^2\Theta}{2}\left(\sqrt{1 + \frac{\varepsilon_{ik}^2}{(\varepsilon_{rk} - \sin^2\Theta)^2}} - 1\right)}$$

(5.10)

$$\tilde{\chi}_k = \sqrt{\frac{\varepsilon_{rk} - \sin^2\Theta}{2}\left(\sqrt{1 + \frac{\varepsilon_{ik}^2}{(\varepsilon_{rk} - \sin^2\Theta)^2}} - 1\right)}$$

The Fresnel coefficients of reflection, with specifications made by Bogorodsky et al. (1976) were calculated using the formula:

$$\left|R_{j,k}\right|^2 = \frac{(\gamma_j - \gamma_k)^2 + (\delta_j - \delta_k)^2}{(\gamma_j + \gamma_k)^2 + (\delta_j + \delta_k)^2}$$

(5.11)

The results of calculations of the angular structure of emissivity for different types of sea ice at $\lambda = 3.2$ cm with different ice temperatures are given in Table 5.1. The values of the constants shown in Figs 4.5(a)–(e) were used in calculations as the initial ones. As is seen, the substantial differences in the emitting properties of different types of sea ice open up prospects for the microwave sounding technique. The ice temperature variations

Table 5.1. Coefficients of the sea ice emission Σ ($\lambda = 3.2$ cm)

	$\Theta°$	0°			20°			40°		
Structure	T_S (°C)	−10°	−20°	−30°	−10°	−20°	−30°	−10°	−20°	−30°
Columnar	V	0.797	0.853	0.911	0.816	0.869	0.923	0.876	0.918	0.958
	H	0.797	0.853	0.911	0.777	0.835	0.898	0.706	0.773	0.849
Frazil	V	0.801	0.855	0.921	0.820	0.871	0.931	0.879	0.920	0.964
	H	0.801	0.855	0.921	0.781	0837	0.909	0.711	0.775	0.863
Multi-	V	0.942	0.948	0.953	0.951	0.956	0.960	0.976	0.980	0.982
year	H	0.942	0.948	0.953	0.933	0.940	0.945	0.895	0.904	0.911

tell least on the emissivity of multi-year ice and manifest themselves most in the emitting properties of frazil ice.

Variations in the microwave emission of sea ice from the data of numerical experiments have been studied in detail by Nikitin (1980), who assumed the thickness of the sea ice skin-layer to be much below that of the fresh-water ice and less than the ice thickness. With bottom layers of sea ice assumed to be water-saturated and coordinating the process of the microwave emission transport, the effective temperature of sea ice was calculated without account of the effect of water and the state of the ice–water interface:

$$T_{\text{eff}} = \int_0^H T(z)\alpha(z)g(z)\exp\left[-\int_0^z \alpha(z')g(z')\,dz'\right]dz \tag{5.12}$$

where

$$g = \left[(n^2 - \chi^2 - \sin^2\Theta)/2\chi^2\right]^{0.5}\left\{1 + (2n\chi)^2/(n^2 - \chi^2 - \sin^2\Theta)^{0.5} - 1\right\}^{0.5}$$

$$\tag{5.13}$$

$\alpha = 2(2\pi/\lambda)\chi$ is the coefficient of absorption for power, $\chi = I_m\sqrt{\varepsilon}$ is the imaginary part of the complex index of refraction, T is the thermodynamic ice temperature, Θ is the incidence angle, and H is the thickness of the effective emitting layer.

According to Bogorodsky and Khokhlov (1978) for salt ice the k can be calculated with high accuracy from the value of attenuation $8.68k = \gamma\lambda/2\pi$. In calculations of T_{eff} for multi-year ice, the real profile of density from the data by Khokhlov (1978) has taken into account, the expression for T_{eff} is written as

$$T_{\text{eff}} = \int_0^H T(z)\alpha(z)g(z)\exp\left[-\int_0^z \beta(z')g(z')\,dz\right]dz \tag{5.14}$$

where $\beta(z)$ is the effective index of attenuation, H is the thickness of the scattering layer, $f \geq 0.1$ is the coefficient of diffusion.

For $z > H_S$

$$\beta(z) = \alpha(z) = 2\frac{2\pi}{\lambda}k[T(z), S(z)] \tag{5.15}$$

For $z \leqslant H_S$

$$\alpha(z) = \alpha[r(x)]$$

$$g(z) = g[r(z)] \tag{5.16}$$

$$\beta(z) = \beta[r(z)]$$

As has been mentioned above, according to Shulgina (1975), the T_B of ice was calculated at the product of T_{eff} by the coefficient of transmission (emission) of the ice–air interface, q. The coefficient q in general depends also on the characteristics of the sub-surface layer, in accordance with Brekhovskikh (1978), for $4\pi[\varepsilon(\Theta) - \sin^2\Theta]1.5/3\lambda \gg [\delta\varepsilon/\delta z]$ the Fresnel formula can be used. For some age levels the snow layer on the ice surface was also taken into account, as well as the ice roughness (outside the pack-ice areas) taken from the observational data on the arctic seas roughness characteristics in the approximation of geometric optics, according to Rytov et al. (1978). Fig. 5.1 shows the angular change of the transmission coefficients q for the springtime white ice for the smooth, rough and rough-snow-covered surfaces

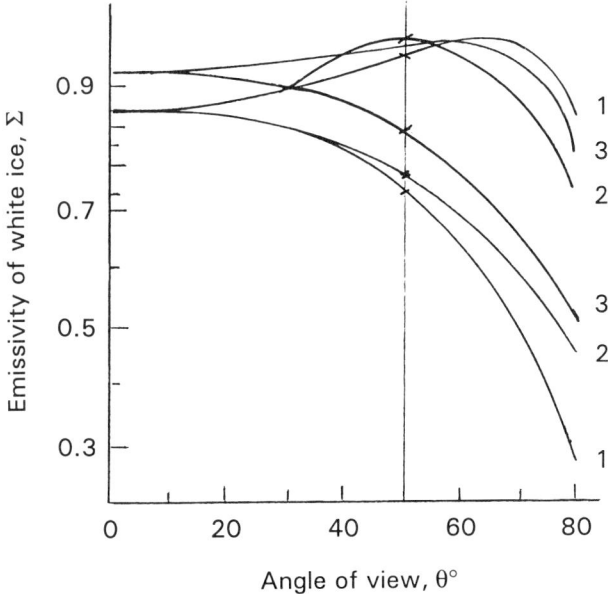

Fig. 5.1. Angular dependence of the emissivity of white ice ($\lambda = 1.6$ cm, vertical and horizontal polarization); 1, smooth surface 2, rough surface; 3, rough surface with the snow cover (Nikitin, 1980)

($\lambda = 1.6$ cm). The calculations show that the roughness of drifting ice manifest itself only at the incidence angles $< 50°$, the snow cover contributes to the T_B to 10 K, the greater the electrodynamic contrast at the snow–ice interface, the greater the contribution (i.e. it is greater for initial, more salty ice forms). Fig. 5.2 shows the results of model calculations (Nikitin, 1980) of the spectral dependence of the T_B of sea ice for the nadir sounding. The observational data obtained by Rabinovich and Melentyev (1970), Campbell et al. (1978), Wilheit et al. (1972), Tooma et al. (1975), Gloersen et al. (1973) are shown, too. The analysis shows that the spectral variations of the T_B of the system of first-year ice are connected, mainly with the dispersion of their dielectric permeability. The effective T_B for most of the cases considered grows with increasing λ. The coefficient q, as can be seen from Fig. 5.1, has an opposite trend. As a result, the value of spectral variations of T_B is at a maximum in the spring: so, for example, T_B of thin ice without snow at $\lambda = 0.8$ cm in the spring is 23 K higher than at $\lambda = 8.0$ cm.

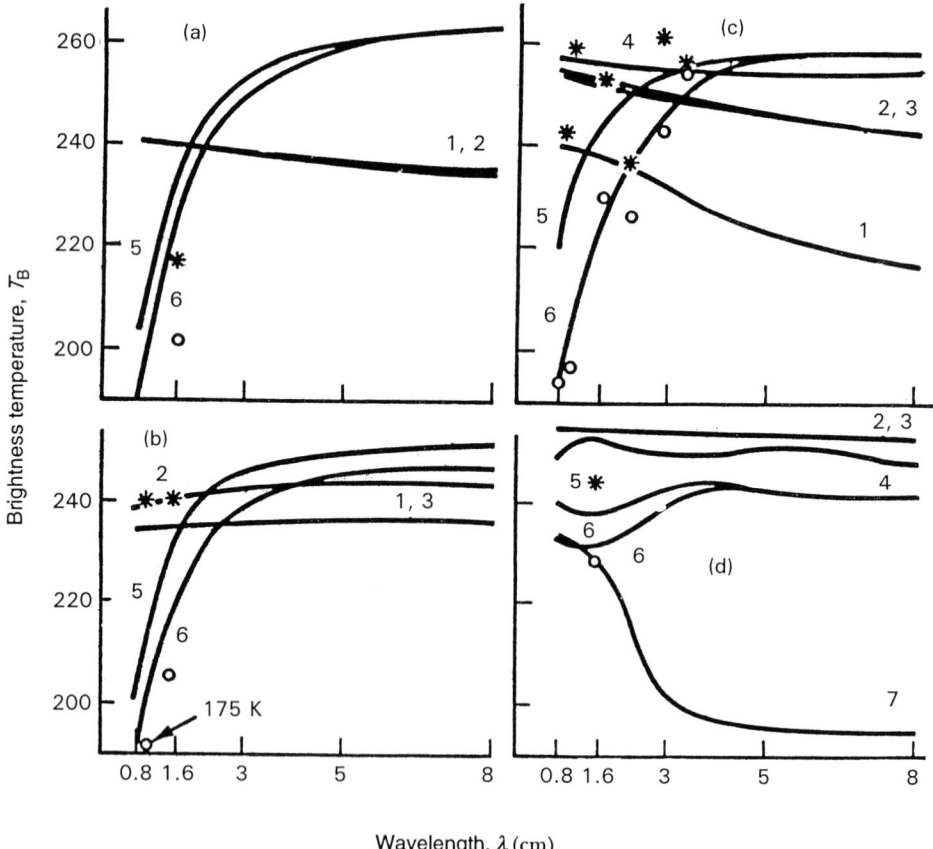

Wavelength, λ (cm)

Fig. 5.2. Brightness temperature of sea ice: (a) autumn, (b) winter, (c) spriung, (d) summer); 1, grey-white ice; 2, white ice; 3, first-year ice; 4, $H > 150$ cm; 5, two-year ice; 6, multi-year ice; 7, wet ice (30%); *, ○, experimental data (Nikitin, 1980)

The multi-year ice is characterized by a monotonous increase of T_B with increasing wavelength (Nikitin, 1980). Such spectral dependence is observed during the whole year, except for the short summer period, when the upper melting snow–ice layer of the arctic drifting ice is water-saturated. However, the decrease of the T_B of the polar multi-year ice can be connected with scattering, too: this effect (Bogorodsky and Khokhlov, 1978; Gloersen et al., 1973) can reach ~25% at $\lambda = 0.8$ cm and stops being seen at $\lambda \geqslant 5.0$ cm. These problems will be considered in detail below. Within the diffuse model (Nikitin, 1980), the T_B difference between multi- and second-year ice turns out to be negligible, despite the substantially different thicknesses of the layers that form the emission.

5.2 TWO-STREAM MODEL OF PURE ABSORPTION FOR THE SEA ICE–SNOW–ATMOSPHERE SYSTEM

Let us consider the results of theoretical calculations performed by Shulgina (MGO, 1978) to assess the contributions of scattering to the emission of sea ice. These data will be useful in further interpretation of aircraft and satellite microwave measurements. In contrast to the previous publication considered and those of some other authors, the expression for the T_B of the three-layer medium "water–ice–air" by Shulgina for the model of pure absorption (without account of the effect of the atmosphere) considers the constituent of the water emission penetrating the water–ice and ice–air interfaces and repeatedly reflected at these interfaces, as well as direct and backscattered (reflected from the ice–water interface) constituents of the ice emission—Fig. 5.3:

$$T_B(\Theta_1) = \left(1 - R_{21}^2\right)\left\{\left[1 + R_{23}^2 \exp\left(-\int_0^{l_2} \tilde{\alpha}\, dz\right)\right]\right.$$

$$\times \int_0^{l_2} \tilde{\alpha}_2 T_2(z) \exp\left(-\int_0^z \tilde{\alpha}_2\, dz'\right) dz$$

$$\times \sum_{n=0}^{\infty} R_{21}^{2n} R_{23}^{2n} \exp\left(-2n\int_0^{l_2} \tilde{\alpha}_2\, dz\right)$$

$$+ T_3\left(1 - R_{23}^2\right)\exp\left(-\int_0^{l_2} \tilde{\alpha}_2\, dz\right)$$

$$\left. \times \sum_{n=0}^{\infty} R_{21}^{2n} R_{23}^{2n} \exp\left(-2n\int_0^{l_2} \tilde{\alpha}_2\, dz\right)\right\} \tag{5.17}$$

where Θ_1 is the viewing angle, T_3, $T_2(z)$ are thermodynamic temperatures of water and ice, respectively, R_{jk} are the Fresnel coefficients of reflection at the interfaces, $\alpha_i = \left(4\pi\chi_i/\lambda\right)$ are the coefficients of absorption of the media, and λ is the wavelength.

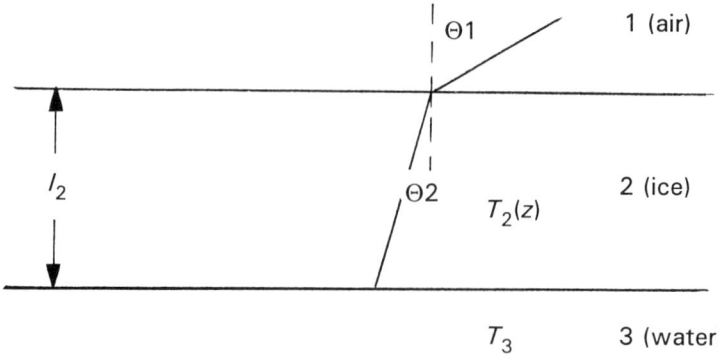

Fig. 5.3. The scheme of emission of the three-layer system (MGO, 1978)

In the case of a four-layer system the emission received by the radiometer for this model consists of three components: water emission penetrating the interfaces water–ice and ice–snow, re-reflected by snow and getting to the air; ice emission (direct and backscattered); and snow emission with account of multiple re-reflections—Fig. 5.4. Formula (5.18) is an expression for the T_B of the four-layer system:

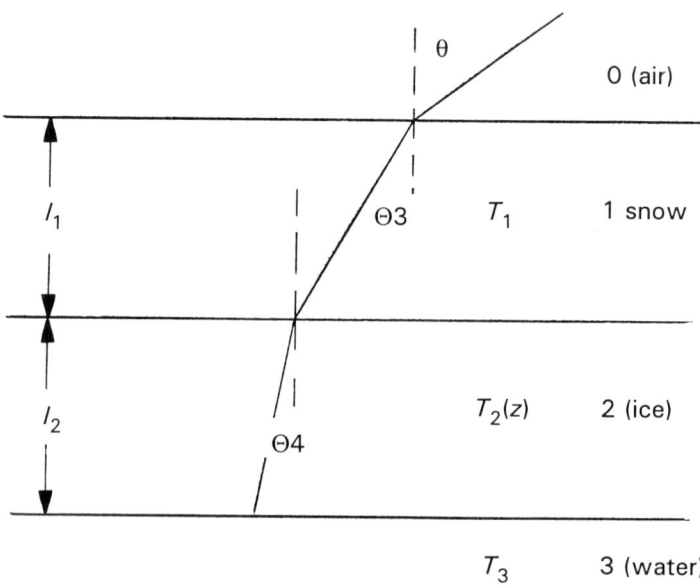

Fig. 5.4. The scheme of emission of the four-layer system (MGO, 1978)

$$T_B(\Theta_1) = \left[1 - R_{10}^2\right]\left\{\left(1 - R_{23}^2\right)\left(1 - R_{12}^2\right)T_3\right.$$

$$\times \exp\left(\int_0^{l_2} \tilde{\alpha}_2 \, dz\right)\exp\left(-\tilde{\alpha}_1 l_1\right)$$

$$\times \sum_{n=0}^{\infty} R_{12}^{2n} R_{23}^{2n} \exp\left(-2n\int_0^{l_2} \tilde{\alpha}_2 \, dz\right)$$

$$\times \sum_{n=0}^{\infty} R_{10}^{2n} R_{12}^{2n} \exp\left(-2n\tilde{\alpha}_1 l_1\right)$$

$$+ \left(1 - R_{12}\right)^2 \int_0^{l_2} \tilde{\alpha}_2 T_2(z) \exp\left(-\int_0^{l_2} \tilde{\alpha}_2 \, dz'\right) dz$$

$$\times \left[1 + R_{23}\exp\left(-\int_0^{l_2} \tilde{\alpha}_2 \, dz\right)\right]\exp\left(-\tilde{\alpha}_1 l_1\right)$$

$$\times \sum_{n=0}^{\infty} R_{10}^{2n} R_{12}^{2n} \exp\left(-2n\tilde{\alpha}_1 l_1\right)$$

$$\times \sum_{n=0}^{\infty} R_{12}^{2n} R_{23}^{2n} \exp\left(-2n\int_0^{l_2} \tilde{\alpha}_2 \, dz\right)$$

$$+ T_2\left(1 - R_{12}^2\right)\exp\left(-\tilde{\alpha}_1 l_1\right)\left[1 - \exp\left(-\tilde{\alpha}_1 l_1\right)\right]$$

$$\left. \times \sum_{n=0}^{\infty} R_{12}^{2n} R_{20}^{2n} \exp\left(-2n\tilde{\alpha}_1 l_1\right)\right\} \tag{5.18}$$

Notation is as in (5.19).

The results of calculations made in accordance with the distribution of specific absorption and refraction index for real distributions of temperature and salinity in each type of ice are given in Table 5.2.

An analysis of calculated data shows that for all types of ice considered the T_B grows with increasing wavelength, which can be explained by the growth of the effectively emitting layer, resulting in the effect of warmer near-to-water ice layers (see Figs 4.9 and 4.10). The two-year ice has a maximum of T_B 258–262 K, depending on the wavelength and it has low coefficients of refraction and absorption and, hence, greater emissivity and maximum thickness of the effectively emitting layer. The value of T_B white ice ranges from 232 to 238 K, depending on wavelength, and wintertime ice has values between 225 and 229 K. Low T_B values for the wintertime ice lacking spectral dependence at short wavelengths (0.72–3.0 cm) are explained by their inherent temperature lapse rate in the upper emitting layers of this ice. At $\lambda = 20$ cm the wintertime ice emission increases

Table 5.2. The brightness temperature (T_B) and thickness of effective emitting layer h_{eff} for different types of ice

Type:	White ice				Winter ice				Second-year ice			
λ (cm):	0.72	1.6	3	20	0.72	1.7	3	20	0.72	1.6	3	20
$T_{B,K}$	229	232	233	236	225	225	225	229	258	259	260	262
h_{eff}, cm	20	20	30	80	10	30	40	140	60	70	90	150
Structure + snow 30 cm thick												
$T_{B,K}$	237	238	239	241	232	232	231	234	258	261	261	263
Structure + snow 70 cm thick												
$T_{B,K}$	238	239	239	241	234	233	232	234	—	—	—	—

negligibly ($\Delta T_B = 4$ K) due to the effect of warm bottom layers. The more strongly expressed spectral dependence of the white ice emission and higher values of the white ice T_B are explained by great values of temperature gradient in the ice thickness. Thus, the dominating effect on the emitting properties of sea ice is exhibited by the vertical temperature distribution inherent to each structure considered.

The effect of snow cover is more apparent in the case of white ice and winter ice, the positive contrast in the presence of snow cover reaching 5–6 K. The presence of snow cover up to 30 cm thick over the two-layer ice practically does not show in the emission characteristics—the T_B contrast does not exceed 1 K.

5.3 COMPARISON BETWEEN THEORY AND OBSERVATIONS, A TECHNIQUE TO TAKE INTO ACCOUNT SCATTERING BY SEA ICE

The emitting properties of sea ice considered above have been calculated for the model of pure absorption. Nevertheless, a comparison of calculation results with those of field experiments in some studies carried out in different countries (for example, Brekhovskikh (1978); Tsang and Kong (1975); Gloersen et al. (1973)) reveals a systematic difference between observations and theory in the case of aircraft and satellite measurements of the microwave emission of multi-year ice. The assumption that this had been caused by neglecting the contribution of scattering was made in MGO (1978).

The multi-year sea ice has in its upper part a considerably thick freshened layer, whose weak salinity, on account of strongly inhomogeneous ice, can cause an effect of additional scattering of microwave emission. Calculations made in the publication mentioned above for real parameters of ice, were made in a two-stream approximation for a semi-infinite stratified structure with the varying temperature profile, with both absorption and scattering taken into account. The results of calculations are given in Table 5.3.

The first column of figures (I) gives the T_B calculated for real temperature profiles and dielectric constants obtained from the data of measurements made with account of the model of pure absorption; the second column (II) is the calculations with a two-stream

Table 5.3. Calculated values of the sea ice brightness temperature T_B

Type:	White ice			Winter ice			Second-year ice		
Technique:	I	II	III	I	II	III	I	II	III
λ (cm)									
0.72	229	226	226	225	225	225	258	258	251
1.25	—	228	227	—	226	225	—	259	250
1.6	232	229	228	225	226	225	259	259	251
3.0	233	230	229	225	226	225	260	260	253
8.0	—	233	232	—	227	226	—	261	258
20.0	236	236	236	229	229	229	262	262	261

model without scattering; the third column (III) is the same but taking account of scattering.

The analysis of the Table 5.3 data shows that for the winter and second-year ice the results of calculations with both models taking into account only the absorption, practically coincide. In the case of white ice, there is a difference, probably connected with large gradients of temperature and dielectric constants with a relatively small (compared to the types of ice mentioned above) thickness of this ice. A systematic decrease of the calculated values of T_B for the two-stream model is connected with limitations of the method itself, which holds only for the cases of weakly inhomogeneous media.

The effect of scattering (columns II and III in comparison) is negligible, according to calculations: at the shortest wavelength 0.72 cm for the two-year ice it reaches a maximum of 7 K. Thus, for all types of sea ice considered, none of the substantial effects, connected with scattering, have been observed either in absolute values of T_B or in the spectral change. Even for the two-year ice, where the scattering process is most substantial, the character of the spectral dependence of T_B on λ practically does not change. For the model of pure absorption, the T_B grows slowly, with the different between T_B at $\lambda = 0.72$ cm and 20.0 cm being only 10 K. For the winter ice the change in T_B at these two wavelengths constitutes only 4 K.

For the white ice, the T_B contrast at two extreme wavelengths considered above is stronger than for the winter ice, but this is connected not with the effect of scattering (columns II and III, respectively) but with the increasing vertical gradient of the thermodynamic temperature for the given type of ice—Fig. 4.3.

Now we shall discuss the contrasting results of calculations and observations obtained by Russian and American experts within the experiments AIJEX and "Bering Sea Experiment". From the data of Ramseier et al. (1975), the measurements within the AIJEX programme in the Arctic Basin at $\lambda = 1.55$ cm gave the T_B for the one-year and multi-year ice 255 and 230 K, respectively. The substantial difference in the T_B is explained by the effect of scattering in multi-year ice. Nevertheless, the results of calculations for the two-stream model described above reveal a decrease of T_B in the whole spectral region from 0.72 to 20.0 cm due to the effect of scattering and give quite opposite data on the variability of the first-year and multi-year T_B at the comparable wavelength 1.6 cm: 225–232 K and 253–259 K, respectively (the scattering of values within each of the age intervals is determined by the model chosen).

To verify the correct choice of the calculation scheme, it was tested (MGO, 1978) with the data on the emissivity of fresh-water ice, where the effect of scattering due to much lower absorption of microwave emission, as compared to sea ice, should manifest itself more strongly. Table 5.4 gives the T_B values calculated for the two-stream model for the fresh-water ice in the case of pure absorption (column I) and in the presence of scattering (column II). Thus, the calculations verified the correctness of the scheme chosen and showed the substantial contribution of scattering for fresh-water ice, for which, at short λ, the decrease of T_B due to scattering reaches almost 30 K, and confirm the prospects for multi-spectral microwave sensing of the ice cover of the inland water basins and continental ice.

Table 5.4. Calculated brightness temperature (T_B) for fresh-water ice

λ (cm)	0.72	1.25	1.6	3.0	8.0	20.0
I—absorption	253	253	253	251	227	181
II—scattering	224	213	208	201	202	176

5.4 EFFECTS OF WATER FILMS ON SEA ICE

Before discussing the problem of microwave sounding of the fresh-water ice of lakes and water reservoirs, we shall dwell upon the problem we have faced in discussing the results of joint studies performed within the programme "Bering Sea Experiment"—the emitting properties at initial stages of newly formed one-year sea ice (Gloersen et al., 1975a). The American scientists revealed the possibility of substantial "brightenings" of the T_B measured by the in-flight radiometer at $\lambda = 1.55$ cm in the region adjacent to the fresh ice–water interface. From the data of Convair-990 aircraft, in these regions the T_B of the thin grey ice corresponds to the multi-year ice. Similar phenomena were observed by the American scientists in the analysis of the materials of the field experiment AIJEX. Note, that from our airborne measurements made in various spectral regions and over various water basins, such an effect was also observed in a number of cases: so, for example, in studies of the processes of initial ice formation (Bespalova et al., 1976) the microwave measurements at $\lambda = 3.0$ cm revealed a considerable increase of the T_B recorded as individual peaks in the regions of polynyas and small-broken ice, as compared to the smoothed profiles of T_B, corresponding to the fields of sufficiently homogeneous sea ice or solid large-broken ice—Fig. 5.5.

In connection with the importance of this problem, since it can limit the possibilities of retrieving the state of sea ice, model calculations were made (Rabinovich et al., 1975) of the emissivity of white, grey-white and nilas, i.e., for all ice types observed in the "Bering Sea Experiment". These types of ice differ in thickness, distribution of dielectric

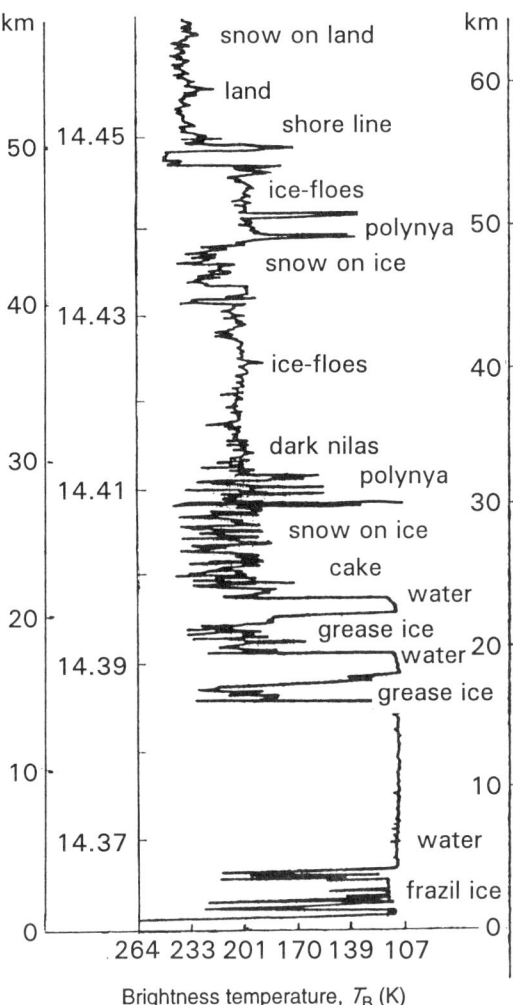

Fig. 5.5. Brightness temperature of the sea ice (Caspian Sea: $H = 500$ m) (Bespalova et al., 1976)

constants and temperature: the corresponding distribution of physical parameters and electro-dynamic characteristics were obtained at the ship "Priboy" simultaneously with the airborne survey (Bogorodsky and Khokhlov, 1975; Vinogradov et al., 1975)—Table 5.5.

The calculations have shown that in connection with the considerable absorption, the effective emitting layer of these types of ice is small: for white ice it is 10–15 cm, for nilas –1–2 cm. The emission of these types of ice seems not to give any information about the thickness of the ice itself; however, the relationship between the surface temperature of ice and its thickness makes useful the remote sensing of the thickness of first-year sea ice. Fig. 5.6 shows the angular dependences of the emissivity of these types of ice. White ice is characterized by maximum emissivity, the youngest is of maximum

Table **5.5.** Electrophysical parameters of the
Bering Sea ice cover, at $\lambda = 3.0$ cm

Type of ice	n	χ
White	1.6–2.1	0.02–0.20
Grey-white	1.7–2.2	0.02–0.20
Grey	1.9–2.2	0.09–0.25
Nilas	2.0–2.2	0.20–0.60

salinity and has the lowest emissivity. As the ice layer grows, the salinity of the upper emitting layers decreases, because the brine flows down to the lower layers, which leads to a decrease of electrophysical constants and increase of emission. The values of grey-white ice shown in Fig. 5.6 are close to the white ice emissivity, and the Σ of grey ice varies at $\Theta = 0°$ from a maximum for nilas to 0.91.

Table 5.6 gives the T_B for seven types of ice calculated from the data on the sea ice constants obtained by Bogorodsky and Khokhlov (1975) (nadir viewing, the height of the receiver position 8.0 km, the cloud-free atmosphere). As can be seen, the increasing trend prevails, on the whole, in the T_B of the ice cover with increasing thickness. However, the variations of T_B within one age interval can be substantial, overlapping the values for the adjacent interval.   The results of calculations of the effect of water films on the ice of the nilas type (this phenomenon can often be seen with relatively high air temperatures in the case of dark nilas, and in the storm conditions at lower temperatures) are shown in Table 5.7.

The calculations show that the presence of water film on the surface of nilas leads to a decrease of the T_B of this ice, so, with small thicknesses of films (< 5 mm) their T_B are even below the temperature of the open sea surface, equal to 105 K, in the same conditions. It should also be stated that the calculations for simple three-layer models fail to simulate the observational data.

The best agreement between the microwave signatures and physical parameters of sea ice is observed for an important geophysical parameter, such as ice concentration. First studies on the remote sensing of ice performed in the Soviet Union (Rabinovich et al., 1970) revealed a linear relationship between the concentration in points or percentage and

Table **5.6.** Calculated T_B for the first-year sea ice

λ (cm):	White ice I	White ice II	Grey-white ice I	Grey-white ice II	Nilas I	Nilas II	Nilas III
0.8	259.1	252.0	260.1	254.4	242.5	236.4	201.4
1.6	258.8	251.1	259.8	253.8	240.8	234.2	196.6
3.2	258.6	250.8	259.7	253.3	240.6	233.7	195.3

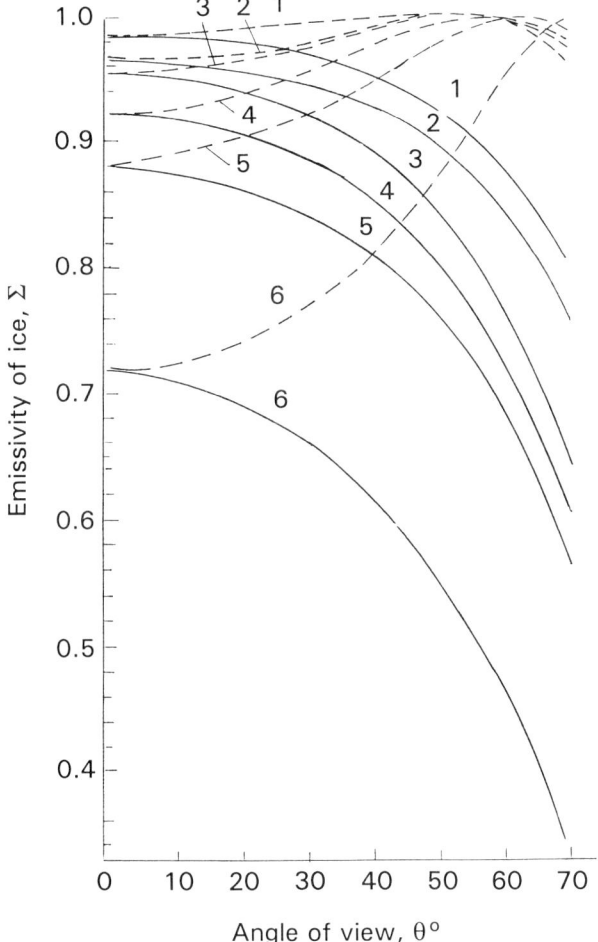

Fig. 5.6. Angular dependencies of the different (1 to 6) ice-types emissivities for the vertical (dashed line) and horizontal (solid line) polarizations. (1, 2) snow over white ice; (3, 4) white ice; (5, 6) nilas (Rabinovich et al., 1975)

Table 5.7. The effect of water film on the emission of nilas
($T_W = T_{air} = 273$ K, $S = 40\%o$, $T_B = 105$ K)

Water film thickness (mm)	0.1	0.3	0.7	1.0	2.0	5.0
T_B ($\lambda = 3.2$ cm)	97.2	97.5	99.1	100.4	104.4	105.0

T_B at $\lambda = 3.2$ cm. The stronger the homogeneity of the ice under study, the less was the deviation from the liner law. Fig. 5.7 shows the dependence of the T_B of the first-year ice of the Bering Sea on the concentration. The scattering of the points averaged over one-minute intervals of the Ilyushin-18 flight turns out to be considerable. The left-hand straight line on the plot corresponds to average T_B values for nilas and grey ice, the right-hand line to those for grey-white ice and white ice. The method of histograms was used to select the prevailing types of ice (Melentyev and Alexandrov, 1989). A similar approach was developed by the American experts (Ramseier et al., 1975) in studies performed within the programme of the "Bering Sea Experiment"—Fig. 5.8. The division of the whole area of survey for Convair-990 into a hundred of observation sectors made it possible to identify grey ice and thick one-year ice over the whole spectral region from 0.8 to 21.0 cm, both on horizontal and vertical polarizations.

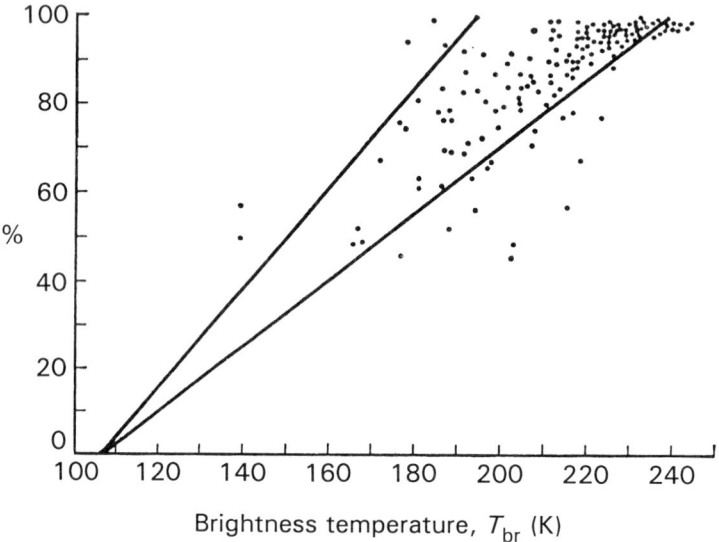

Fig. 5.7. Brightness temperature versus first-year ice concentration (Rabinovich et al., 1975)

5.5 MICROWAVE PROPERTIES OF ICE COVER OF INLAND BASINS AND CONTINENTAL ICE

Calculations of emissivity of lake and river fresh ice as well as frozen soils were made by Kondratyev et al. (1989) for aims to interpret the results of remote sensing. The data on the constants mentioned above and the data from a study by Ray (1972) were used as the initial ones. The calculations were made for the following informative wavelengths: 0.8, 2.5, 11, 18, 30, and 60 cm on horizontal and vertical polarizations, the ice thicknesses were up to 140 cm, and viewing angles varied from 0° to 60°.

Fig. 5.9 illustrates the dependence of emissivity on the ice thickness for nadir viewing at the ice temperature –40°C. As seen from the plot, studies of the fresh-water ice up to 20 cm thick should be made at $\lambda = 0.8$ and 2 cm, for which the contrast $\Delta\Sigma = \Sigma_{20} - \Sigma_0$

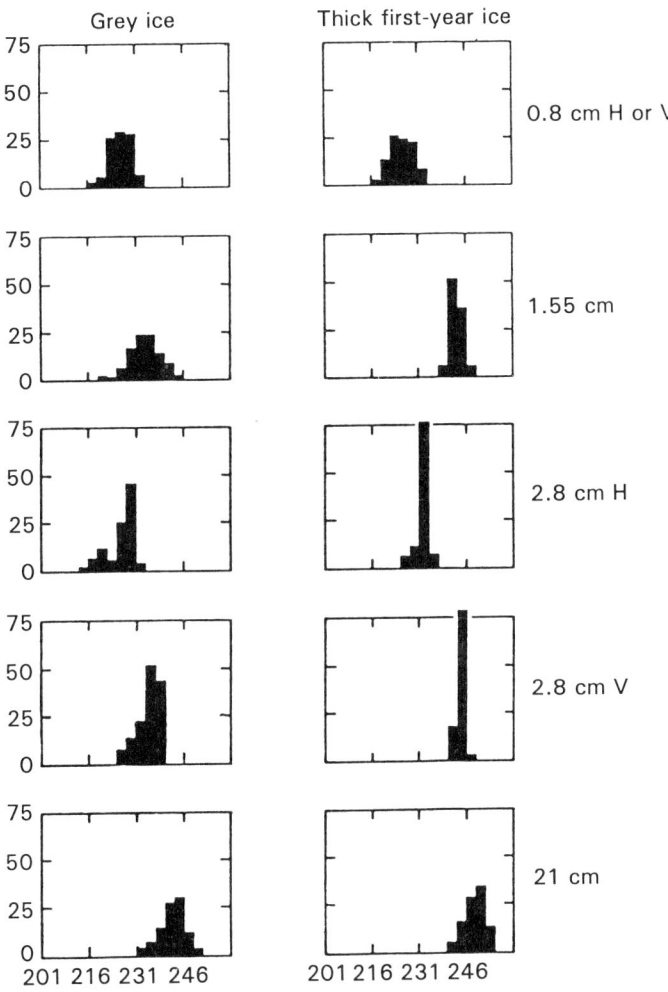

Fig. 5.8. Histograms for different frequencies, polarizations, and ice types, 5 March 1973
(Ramseier et al., 1975)

constitutes, respectively, 0.41 and 0.37. At these wavelengths a highly linear dependence is provided of emissivity on the young ice thickness. For $\lambda = 11$ cm the ice up to 5 cm thick is practically transparent for the microwave region and, then, with the growth of ice formation the dependence becomes monotonically increasing, the linear relationship is observed to the ice thickness 90 cm. For $\lambda = 18$ cm the dependence on thickness is monotonic over the whole range of ice thicknesses. However, for the ice less than 15 cm thick the contrast $\Delta\Sigma$ is very small. It is expedient to develop a two-frequency technique to retrieve the thickness of the fresh-water ice; the recommended pairs of wavelengths 18–0.8 cm and 18–2 cm enable one to identify the thickness of lake and river ice over the whole range of its variations.

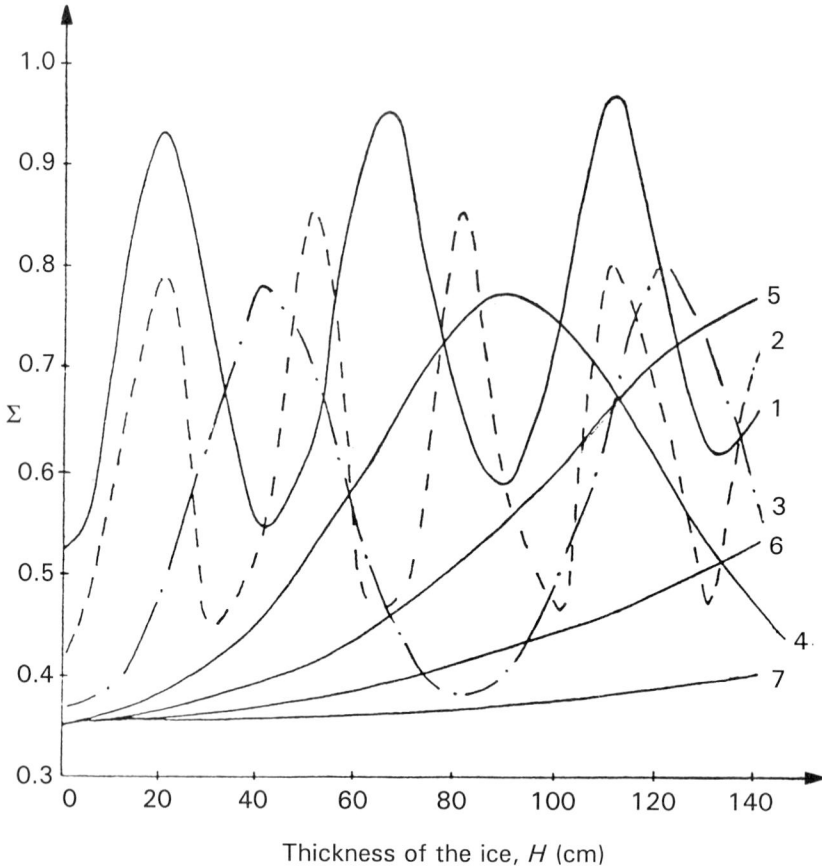

Fig. 5.9. Emissivity of the system: fresh-water–ice–atmosphere versus different ice thickness, $T_{ice} = -40°C$, $\lambda = var$ (1–0.8 cm, 2–2.0 cm, 3–5.0 cm, 4–11.0 cm, 5–18.0 cm, 6–30.0 cm, 7–60.0 cm); $\theta = 0°$ (Melentyev and Aleksandrov, 1991)

Fig. 5.10 presents the dependence of emissivity on the fresh-water ice thickness at various viewing angles on two orthogonal polarizations at $\lambda = 18$ cm. As can be seen from the figure, the deviation of the viewing angle from nadir markedly broadens the possibilities for the sensor with the use of horizontal polarization: the contrast $\Delta\Sigma$ increases greatly with the growing Θ. However, unfortunately, it worsens the linear relationship between T_B and thickness as well as the conditions for identifying ice thinner than 20 cm. On the vertical polarization the character of the angular dependence of emissivity of the three-layer medium "water–ice–air" is complicated: here the increase of the viewing angle Θ worsens the possibilities of retrieving the fresh-water ice thickness (the contrast decreases). The calculations reveal an interesting thing: for $\Theta = 60°$ the Σ practically is independent of ice thickness. This fact opens up possibilities of using the data from microwave sounding to retrieve the thermodynamic temperature of the ice surface. The observed peculiarity of ice emission can be also used to retrieve the SAT from the data of microwave sounding.

Let us consider the very important (from the viewpoint of interpretation) polarization properties of fresh-water ice based on the data from Kondratyev et al. (1992b). Table 5.8 illustrates the calculated spectral-angular dependences at the ice temperature –20°C and ice thickness 10 cm. Fig. 5.11 shows the angular distributions of Σ at $\lambda = 18$ cm for the ice thicknesses 10, 50, 100, and 140 cm.

Table 5.8. Emissivities of fresh-water and the three-layer medium "water–fresh-water ice–air"

Θ	Polariz-ation	λ (cm) 0.8	2.0	5.0	11.0	18.0	30.0	60.0
0° (water)	V, H	0.524	0.421	0.364	0.354	0.353	0.353	0.353
0°	V, H	0.720	0.590	0.388	0.360	0.355	0.354	0.353
10°	V	0.728	0.591	0.392	0.364	0.360	0.358	0.357
10°	H	0.721	0.585	0.383	0.355	0.351	0.350	0.349
20°	V	0.752	0.596	0.405	0.377	0.373	0.372	0.371
20°	H	0.725	0.569	0.370	0.342	0.338	0.368	0.336
30°	V	0.786	0.606	0.428	0.401	0.397	0.396	0.395
30°	H	0.733	0.542	0.347	0.321	0.317	0.315	0.14
40°	V	0.817	0.623	0.464	0.439	0.436	0.434	0.434
40°	H	0.745	0.501	0.314	0.290	0.286	0.285	0.284
50°	V	0.828	0.648	0.520	0.498	0.495	0.494	0.493
50°	H	0.760	0.444	0.271	0.250	0.246	0.245	0.244
60°	V	0793	0.678	0.602	0.587	0.585	0.585	0.584
60°	H	0.767	0.368	0.218	0.200	0.198	0.197	0.196

It can be seen from the data that for fresh-water ice the character of the angular dependence is determined by dielectric properties of all the three media, and that the lake and river ice up to 50 cm thick is sufficiently transparent to the flux of the water's own emission, which determines the character of the angular dependence of Σ. For extreme values of the thicknesses of lake ice 10–140 cm the coefficient of polarization of the medium decreases, the character of the dependence of Σ on Θ both on vertical and horizontal polarizations is mainly determined by the dielectric properties of the fresh-water ice itself. The revealed peculiarities of the emitting properties should be taken into account in designing microwave radiometers: the scanning normal to the flight direction it is still possible to retrieve the characteristics of the state of the three-layer medium "water–ice–air". Fig. 5.11 shows the values of Σ for different temperatures of ice. The increase of the temperature of the air and ice to –10°C raises the Σ for all ice thicknesses and every sounding condition. The increase of the Σ on both polarizations shows itself stronger with ice thicknesses at a maximum.

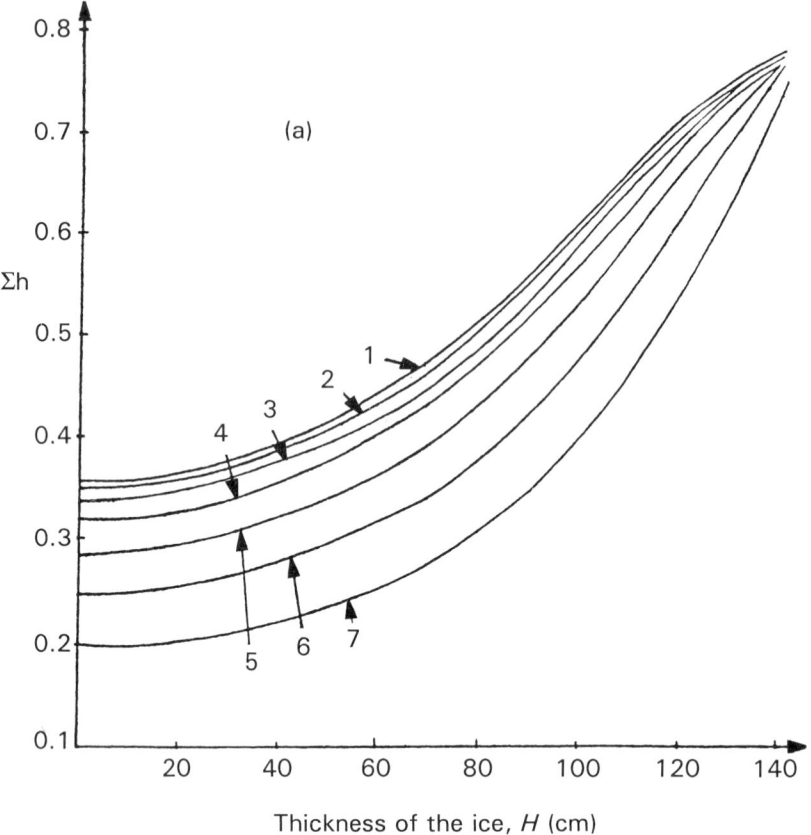

Fig. 5.10. Emissivity of the system: fresh-water–ice–atmosphere versus different ice thickness, $T_{ice} = -10°C$; $\lambda = 18.0$ cm; $\theta =$ var (1–0°, 2–10°, 3–20°, 4–30°, 5–40°, 6–60°, 7–70°). (a) Horizontal polarization; (b) vertical polarization (Melentyev and Aleksandrov, 1991)

The calculations of the microwave emission of the extended layer of the continental ice up to 1000 m thick showed (Melentyev and Alexandrov, 1991) small interference oscillations of Σ about 0.02 for wavelengths 30 and 60 cm. At the same time, apparently, for such cases it is necessary to take into account the volume scattering since the calculated values of Σ are overestimated (approaching unity), which contradicts the data of the Ilyushin-18 MGO experiments over the glaciers of Novaya Zemlya.

The observational studies of the emitting properties of real soils and water medium in its various states revealed the interference effects (Tuchkov, 1968), which further confirm that the model of transfer of the emission of a flat-stratified isothermal structure with sharp interfaces between the media has a physical sense. The calculations made for the Σ of fresh-water ice should be supplemented with taking into account the presence of air bubbles in the fresh-water ice. The analysis of the lake ice cores shows that maximum ice porosity, for example, of Onega Lake constitutes 3%, the size of air bubbles is 1–2 mm. Calculations of the effective dielectric permeability of this ice were made with the use of

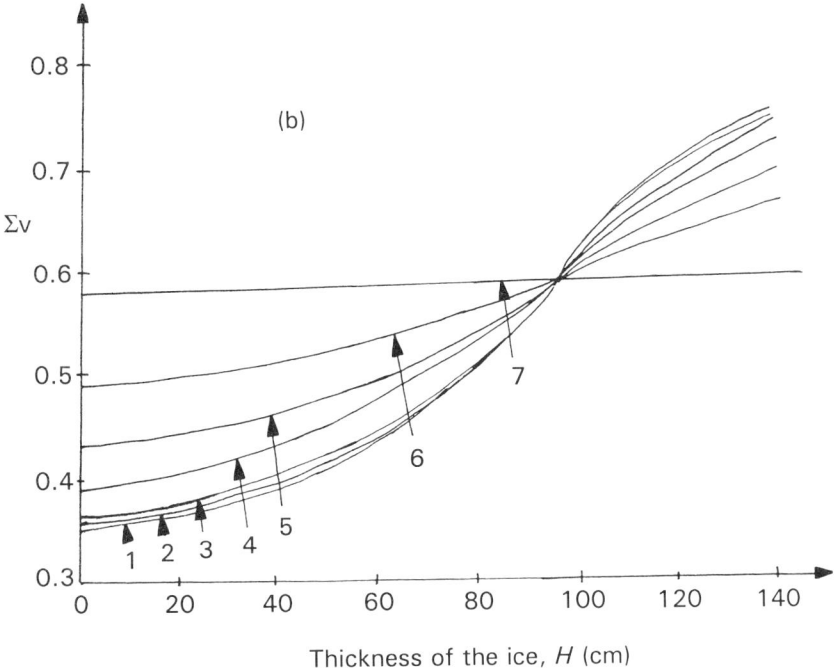

Fig. 5.10(b).

the Weiner equation in the complex form. Fig. 5.12 shows the dependence of the difference between the emissivities of the porous ice, with the value of porosity 0.03, and
monolithic ice on the ice thickness. At $\lambda = 10$ cm the air bubbles slightly reduce the Σ
(by 0.02 at a maximum), and at longer wavelengths their effect, in practice, does not
show. At the same time, in the centimetre range the air bubbles vary substantially the
interference pattern, either increasing or decreasing the Σ for different thicknesses of ice.
At $\lambda = 0.8$ cm the $\Delta\Sigma$ reaches 0.175, at $\lambda = 2$–0.08 cm, and then, with the wavelength
increasing, it decreases more substantially. Naturally, the effect of air bubbles is at a
maximum for thicker lake ice.

Now we shall examine the data of calculations of the emissivity of frozen soils—Fig.
5.13. This sphere of application of microwave data is useful to study the problem of the
arctic climate, and is still in its initial stage (publications on this problem are scarce). Few
observational studies are known (Arcorn et al., 1979; England, 1979; Olhoert, 1978) the
principal conclusion from which is that the frozen soil is radiometrically dry, and the line
of freezing divides the frozen and unfrozen soils as media with contrasting dielectric
properties. As can be seen from Fig. 5.13, in nadir viewing the effect of snow over the
frozen soil is the strongest in the shortwave interval of the microwave region. The dry
snow 10 cm thick raises the emissivity of the frozen clay at $\lambda = 0.8$ cm from 0.871 to
0.934, which in T_B constitutes about 16 K. However, in this part of the spectral region the
oscillations are also at a maximum, therefore, longer wavelengths can be recommended
to retrieve the snow thickness: so, for example, for snow thicknesses less than 60–70 cm

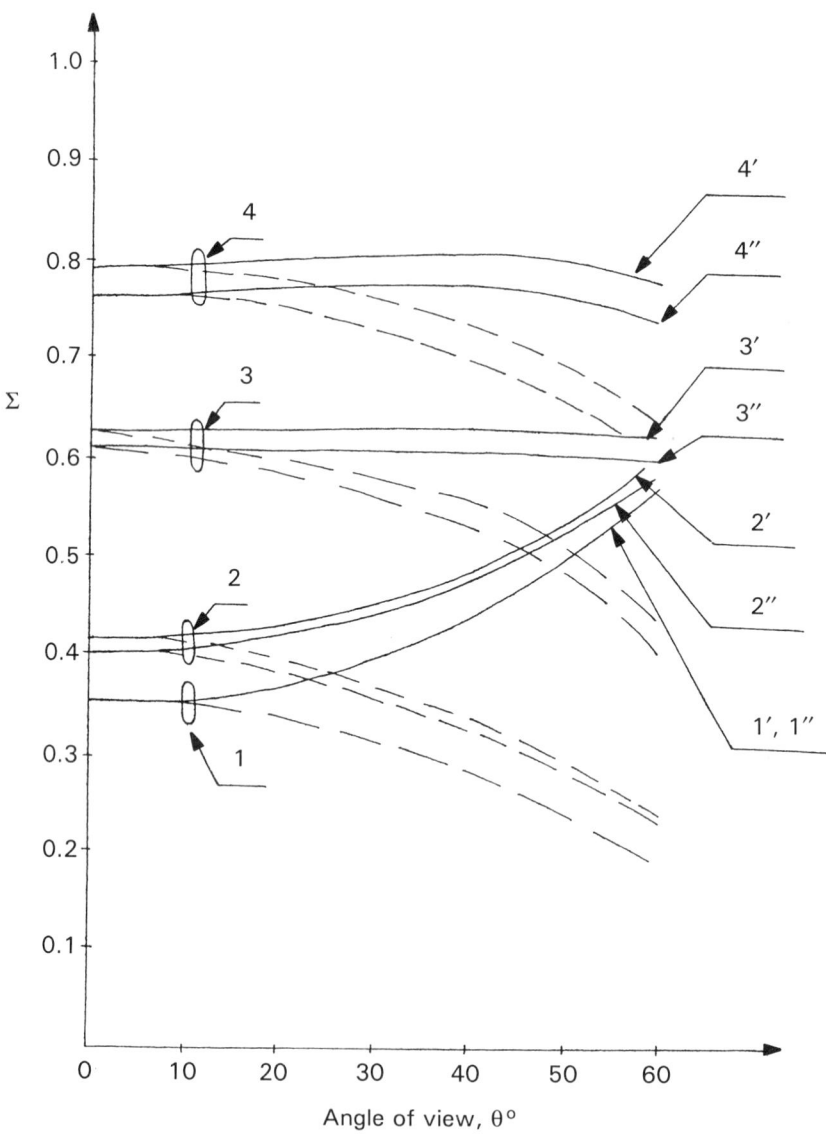

Fig. 5.11. Emissivity of the system: fresh-water–ice–atmosphere versus angle of view, $T'_{ice} = -$ 10°C; $T''_{ice} = -40°C$, $\lambda = 18.0$ cm; $H_{ice} = var$ (1–10 cm, 2–50 cm, 3–100 cm, 4–140 cm). Solid line, vertical polarization, dashed line, horizontal polarization (Melentyev and Aleksandrov, 1991)

at $\lambda = 5$ cm the relationship of Σ with snow thickness is almost linear. At $\lambda = 11$ cm the increase of snow thickness to 110 cm monotonically increases the Σ of the system; however, the contrast, $\Delta\Sigma$, becomes small. At $\lambda = 60$ cm the dry snow of any thickness is almost transparent in the microwave region.

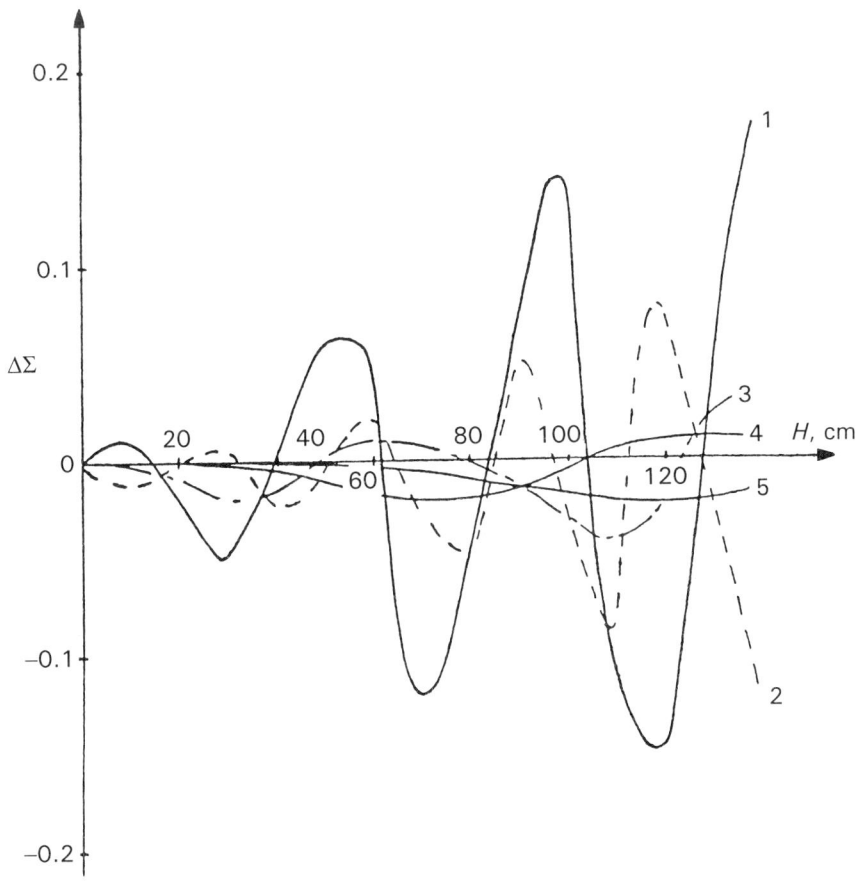

Fig. 5.12. Difference of emissivity ($\Delta\Sigma$) of porous and solid fresh ice ($T_{ice} = -10°C$) (Melentyev and Aleksandrov, 1991)

From the data of calculations the polarization properties of the microwave emission of clay are strongly expressed. Of interest are a significant result of our calculations: they revealed a substantial difference in the microwave emitting properties of the frozen soils and fresh-water ice. This circumstance permits one, using the data of microwave survey, to identify the boundaries of frozen rivers, lakes and tundra marshes in the presence of the fast ice, which is rather urgent in the accomplishment of various geological and geophysical studies in the arctic regions. Such data on the actually existing balance of the arctic surface waters for various climatological problems are also useful. Note that the data of calculations testify to the fact that the thin, newly formed ice on the surface of fresh-water basins can easily be identified using microwave radiometry.

Similar calculations were made for the model "air–frozen sandy loam–alumosilicates". The absolute values of emissivity for such a medium lie within the range 0.76–0.95, which makes it possible to draw the conclusion about the principal possibility of using the microwave survey to identify the types of land surface at negative temperatures.

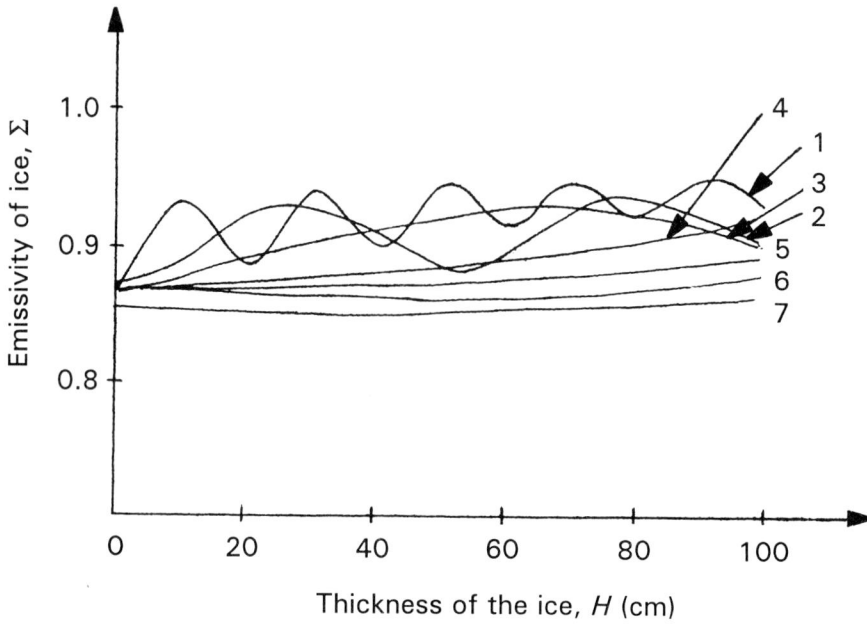

Fig. 5.13. Emissivity of snow with frozen soil, $\theta = 0°$; $T_{snow} = -10°C$; $\lambda = $ var (1–0.8 cm, 2–2.0 cm, 3–5.0 cm, 4–11.0 cm, 5–18.0 cm, 6–30.0 cm, 7–60.0 cm) (Melentyev and Aleksandrov, 1991)

In summary, the calculation data presented here can serve as the basis for the development of techniques to retrieve the characteristics of natural surfaces in the polar regions, of key importance in studies of climate problems. The analysis of the data by various authors reveals the quantitative disagreement between the data of various theoretical studies, which testifies to the need for their further improvement, and the need for complex observational studies in which multi-spectral remote sensing is combined with *in situ* measurement data; the idea of combination of field measurements in conditions of the controlled experiments with aerospace remote sensing remains urgent at the present stage of remote sensing developments. The results of key research programmes, combining such multi-spectral remote sensing and *in situ* measurements in the Arctic, and related research missions by American, European, and Russian researchers are dealt with in Chapters 6 and 7.

6

Problems of climatic interpretation of the passive and active microwave satellite data

6.1 SPECIFIC FEATURES OF THE MULTI-SPECTRAL MICROWAVE SENSING OF POLAR REGIONS

The all-weather nature of microwave remote sensing of surface parameters and the possibility of quantitative interpretation of the results obtained have stimulated the development of techniques to interpret the microwave data. Good prospects for this direction of research have been discovered since the analysis of information of the first "Kosmos-243" microwave satellite launched in the USSR in 1968, and of the following similar satellites (Basharinov and Gurvich, 1970). The satellite survey (scanning system functioned at $\lambda = 3.2$ cm, two orthogonal polarizations) demonstrated that the temporal and spatial distribution of sea ice in arctic and antarctic regions can be determined from the multi-spectral passive microwave data. But recognizing the age of sea ice from the microwave data is still a problem (Kondratyev et al., 1992b).

The first American microwave imager was the Electrically Scanning Microwave Radiometer (ESMR) on board the Nimbus-5 satellite, launched in 1972. The single-channel instrument was a cross-track scanner with incidence angle varying from 0° to 50°, and it recorded horizontally polarized radiation at the wavelength of 1.55 cm. The ESMR data have been used to determine and map global sea ice distribution and concentrations at a grid spacing of approximately 30 km. The resolution of this instrument varied from 30 to 50 km, except the area poleward of 85° N latitude. The ESMR instrument transmitted high-quality data for most of the period 1973–76, and monthly averages of the ice cover have been collected in two atlases, one for the south polar region (Zwally et al., 1983) and one for the north polar region (Parkinson et al., 1987).

The application of the Nimbus-7 Scanning Multichannel Microwave Radiometer (SMMR), launched in 1978, made it possible to obtain microwave images on a global scale. Microwave images of the Earth obtained by Gloersen and Campbell (1988, 1991) are of considerable interest.

The contrast of radio-brightness temperatures (T_B) of land and ocean reaching 160 K are most clearly seen in microwave images. Low T_B over the ocean increase significantly in the presence of clouds or precipitation zones. Therefore the Intertropical Convergence Zone (ITCZ) and extratropical frontal zones are clearly seen. In polar regions, distributions of ice cover are observed, with possible recognition of first-year ice ($T_B \sim 240$ K). The sea ice around the Antarctic is mainly one year old. The multichannel (10 channels) nature of the SSMR data allowed to determine for the wintertime in the Arctic, ice concentrations of two primary ice types: ice that has not undergone a summer melt period, termed "first-year ice", and ice that has undergone a summer melt period, termed "multi-year ice" (Cavalieri et al., 1984). Unfortunately, the Nimbus-7 swath width did not allow SMMR data coverage poleward of about 84.6° N. An atlas presenting monthly averages of sea ice concentrations has been produced by Gloersen at al., (1992). The Atlas includes northern and southern hemisphere monthly average total ice concentration maps for each month from November 1978 through August 1987, plus arctic multi-year ice concentration for the winter months and derived monthly average sea ice temperatures for each hemisphere.

Within the Defense Meteorological Satellite Program (DMSP) was launched a Special Sensor Microwave Imager (SSMI) in 1987 which is still functioning at the time of writing (October 1995). SSMI has seven channels. Plans exist for continuing satellite microwave observations by manufacturing a Multi-Frequency Imaging Microwave Radiometer (MIMR) for the Earth Observing System to be accomplished near the end of 1990s.

Passive microwave data are applicable for long-term, global climatic studies. Satellite active microwave imaging is also providing extremely valuable data for sea ice climatic research. Synthetic Aperture Radar (SAR) has much better spatial resolution than passive microwave radiometers. SAR data can be used for identifying individual ice floes, for recognizing the age of pack and fast-ice structures. SAR instruments have flown on NASA's Seasat (1978), on Russian satellites "Kosmos-1870" (1987), "Almaz" and "Almaz-1" (1991, 1993), on the European Earth Resources Satellite (ERS-1), launched 1991, and the Japanese Earth Resources Satellite (JERS-1), launched in 1992. A SAR instrument with large swath width of 500 km is also planned for the Canadian Radarsat scheduled for launch in 1996.

6.2 POLAR SEA ICE DISTRIBUTIONS: GLOBAL MICROWAVE SURVEY

Now we shall discuss in more detail the results of the interpretation of microwave images, to characterize the spatial distribution of snow and ice cover.

An analysis of the T_B maps for Greenland and the Antarctic for the years of the Nimbus-5 operation illustrated their reliable use in characterizing the ice cover, but did not reveal any marked interannual variations in the emissivity field. Strong seasonal variations were observed in Greenland, connected with melting and freezing. In this connection, an attempt has been made to map the boundaries of snow melting.

The successful operation of SMMR at 1.55 cm carried by Nimbus satellites made it possible to obtain data needed to map the polar ice (the spatial resolution of images was 32 km). Zwally et al. (1983) analysed the microwave images of the northern and southern polar caps for the winter season with available measurement accuracy, the ice cover concentration was estimated with an accuracy of not worse than 6%.

An analysis of a number of sets of microwave images for each polar zone revealed: (i) large differences between the data of various climatic atlases about the ice cover of polar regions and observed distributions of the ice cover; (ii) a marked difference between the distribution of multi-year ice in the north-polar zone and that predicted based on the use of existing models of the ice cover dynamics; (iii) the absence of a non-regular edge of pack ice in the Antarctic; (iv) peculiar contours of T_B isolines for the glaciers of Greenland and the Antarctic, which are determined, apparently, by the morphology of the snow and ice cover, the variability of temperature of the upper layer of ice being an important factor in the Antarctic.

The dynamics of polar ice point to the necessity for satellite monitoring of the varying ice situation. Minimum T_B were recorded in the regions of Greenland (155–200 K) and the Antarctic (130–160 K). In the Antarctic, a good correlation was observed of the fields of T_B and the temperature of the upper layer of the continental ice cover. However, there is no such correlation in Greenland.

Various airborne validation experiments and, in particular, complex sub-satellite measurements with the use of *in situ* observations of the ice properties are very important in developing reliable techniques for the remote sensing of the ice cover properties. For example, the Soviet–American "Bering Sea Experiment" exemplifies as accomplishment of such a programme (Soviet–American..., 1975; USSR/USA..., 1982).

Within the programme of studies of arctic ice dynamics, an airborne microwave and infrared survey was made in the region of the Beaufort Sea from the point with the coordinates 75° N, 150° W, where direct measurements of the ice cover characteristics were made, to 81° N. The results obtained were compared by Gloersen et al. (1973) with the microwave map of the Arctic drawn from the Nimbus-5 data. Five zones of the ice cover with different morphology and dynamics characterized by inhomogeneous T_B fields were found and described in detail. Measurements at wavelengths 0.8 and 1.55 cm have shown that the shortwave microwave emission is very informative from the viewpoint of the characteristic of the ice cover. Though the existence of the five zones mentioned cannot be considered stable, this phenomenon can be supposed typical of the conditions of the Nimbus-5 microwave images.

Campbell et al. (1975, 1976) discussed the efficiency of the passive and active microwave techniques as well as IR images to study sea and lake ice cover—its meso- and macro-structured, and its drifting and deformation, to identify the types of ice and to determine its roughness.

The low resolution of TV images obtained with early meteorological satellites strongly limited the possibility of their glaciological interpretation. The four-channel Landsat images, with a cartographic resolution of about $1:10^6$, have created new possibilities. Examples have been given of the interpretation of these data, which illustrate the potential to identify the ice in the images for all the channels, but channels 0.5–0.6 and 0.6–0.7 µm are most efficient to detect the ice edge and to reveal the structure of ice surfaces.

Images obtained with the NOAA-2 two-channel radiometer were used to map such large-scale inhomogeneities as the pack ice edge, large polynyas, etc. More informative are the data of the two-channel AVHRR. In this case the resolution constitutes about 1 km. The frequent repeatability of the orbit of the NOAA satellite is an important advantage of this data.

Campbell et al. (1975) illustrated the possibility of interpreting images obtained from the four-channel scanning radiometer carried by the DMSP satellite (these images have the resolution in nadir from 0.5 to 3.7 km), as well as photos taken from the Skylab manned orbital station.

Based on the data of the airborne survey and Nimbus microwave images, the prospects for application of passive and active microwave observations were discussed in detail. The airborne microwave survey at 19.3 GHz followed by microwave measurements at frequencies 1.42, 4.99, 10.7, and 37.0 GHz confirmed that data on the ice cover in the presence of clouds and at any time of day could be obtained and made it possible to develop a technique to distinguish between different ice ages. The ground-based measurements showed that differences in emissivity due to ice age were connected with the effect of the surface layer salinity and the size of pores.

The microwave images can be efficiently used to study the morphology and dynamics of the ice cover during 24 hours and in all-weather conditions. The satellite microwave images are particularly important to study the polar ice cover variations and, as has been mentioned above, reveal serious differences with the data of the polar ice atlases. An analysis of images obtained with the side-looking radars and SAR instruments has opened up possibilities to not only reveal morphological details and dynamics of ice, but also to assess the ice cover thickness and hummocking and to more precisely distinguish between the ice types.

The Nimbus-5 satellite also has the microwave spectrometer (MS) on board which has been designed to solve the problems of remote sounding of the Earth's atmosphere: to retrieve the vertical temperature profiles (channels at frequencies 53.65, 54.9, and 58.8 GHz in the 60 GHz band of oxygen), and to retrieve the water vapour and liquid water content in the atmosphere (channels 22.2 and 31.4 GHz). Since for these two channels the polar atmosphere is not totally opaque, the surface contributes significantly to the outgoing emission at the respective frequencies. This makes it possible to use the MS data for measurements in channels 22.2 and 31.4 GHz to obtain information on global distribution of the snow and ice cover. The spatial resolution at the surface level was about 200 by 300 km^2.

As Künzi et al. (1976) showed, it is important for the interpretation of the data considered that the process of internal scattering much affects the microwave emission of most types of ice and snow. The average $T_B = (T_{B,22.2} + T_{B,31.4})/2$ and temperature gradient $\Delta T_B = (T_{B,31.4} - T_{B,22.2})/\Delta u$, are taken as parameters characterizing the distribution of snow and ice cover, the accuracy of absolute T_B measurements being about 2 K, and that of relative ones better than 1 K. Therefore the MSD of absolute values of T_B and ΔT_B are ~2 K and 0.2 K GHz^{-1}, and those of relative values 1 K and 0.1 K GHz^{-1}.

In typical conditions of atmospheric humidity and snow and ice emissivity, corrections of $\overline{T}_B$ and ΔT_B for the effect of the intermediate atmospheric thickness constitute less than 4 K and 0.1–0.1 K GHz^{-1}. respectively. Since these corrections are small as compared to variations of the parameters considered, they were not taken into account.

The ice concentration maps which were produced by NASA experts have been colour-coded for easy visualization. Each colour shade corresponds to an ice concentration increment of 4%. In most instances, the sea ice edge easily identifiable on the maps, as it generally occurs at a fairly sharp transition from the light blue of open ocean to the yellow, browns, and reds of the sea ice cover. The gridded sea ice concentrations has also been integrated to obtain time series of the area of sea ice covered in each of several regions in the Arctic and Antarctic, as well as in the north and south polar regions as a whole. Fig. 6.1 (see the colour section) demonstrates the total sea ice concentration for the winter (February) and summer (August) season in the northern hemisphere (Gloersen et al., 1992).

The passive microwave data reveal that sea ice extent in the northern hemisphere typically ranges from minimum of approximately $8 \times 10^6 \, km^2$ in September to a maximum of approximately $15 \times 10^6 \, km^2$ in March. Regional features of ice distributions were also studied. These data reveal several interesting large-scale climatic phenomena. Strong spatial asymmetry in the wintertime sea ice cover is caused by a variety of geographic, oceanographic, and atmospheric factors. For example, these data permit study of the thermal and ice regime of Barents, Pechora and Kara Seas, to evaluate the influence of the warm Atlantic waters which advected into these seas by the north-flowing Norwegian Current.

Antarctic sea ice climatic variability also was assessed on the basis of microwave data: the sea ice cover of the southern hemisphere, unconstrained by land along its equatorward perimeter, experiences a seasonal cycle having a much greater amplitude than that of the ice cover in the northern hemisphere. In the southern hemisphere, the ice extent typically ranges from a minimum of approximately $4 \times 10^6 \, km^2$ in February, only half the minimum in the northern hemisphere, to a maximum of approximately $20 \times 10^6 \, km^2$ in September, which exceeds by about $5 \times 10^6 \, km^2$ the maximum in the northern hemisphere.

6.3 SNOW COVER GLOBAL DISTRIBUTION: MICROWAVE SATELLITE DATA

As we mentioned above the dielectric constants of water, ice and snow are so different that a weak melting causes a strong emissivity responses (Jones, 1983). The dielectric constant of snow is usually lower than of dry soil and the strong radio-brightness contrast between snow and snow-free surfaces permits the mapping of snowfields using satellite microwave instruments (NASA, 1982).

Künzi et al. (1976) discussed the Nimbus-5 data for the winter and summer seasons in both polar regions, whose analysis testified to the representative nature of average T_B and its gradient in the cases of snow, sea ice and continental ice in Greenland and in the Antarctic, when the snow and ice cover extent is rather great (about or more than $10^5 \, km^2$). Seasonal variations of snow and ice are clearly seen. The use of the two-channel data enables one to identify different types of sea ice and firn. The multi-frequency remote sounding can be used to map the snow cover on the land and sea ice.

An important parameter, characterizing the snow cover, is its water equivalent, knowl-edge of which is particularly important for the forecasts of river run-off (Siberia's river run-off is an important climatic parameter). The earlier results showed that the water equivalent correlates with the snow cover emissivity in the microwave region. An analysis of the data of the microwave emission measurements made with the Nimbus-5 five-channel MS has shown that the data for channels 31.4 and 22.2 GHz can be used to identify the snow cover, since the emissivity of snow at the first of these two frequencies is about 10% less than at the other, whereas the emissivity of bare land surfaces in both cases is the same and constitutes 0.96.

Apparently, the increase of the snow cover emissivity with decreasing frequency is explained by the effect of strong scattering of radiation by inhomogeneities of the upper part of snow cover at higher frequencies, which determines a decrease of emissivity. An increase of emissivity with increasing wavelength can be caused, in the case of dry snow, by increasing thickness of the layer of emission formation: the thickness of this layer can constitute about 1 m, that is, soils can make some contribution to the emission.

The snow cover identification technique mentioned has been used to map the snow cover from the data for various five-day periods from December 1972 to December 1973, with only extended zones of snow cover in Europe, Asia and North America being considered because of the MS low spatial resolution (only 170 km). The computer-drawn maps of the NH snow cover showed that the snow cover extent in December 1972 was much greater than in December 1973. This can be illustrated by the data on the share of territory (percentage) covered with snow for Europe (30–60° E, 50–63° N):

Table 6.1. Average Europe snow extent overland (1972–1973)

1972			1973		
23–28	21–27	15–18	17–21	22–26	22–26
December	January	April	June	August	December
67	59	5	0	0	45

The limited observational data available at the time did not permit one to solve the problem of determination of the snow cover water equivalent. In this connection, ground-based studies have been accomplished of special features of the spectrum of the snow cover microwave emission in the interval 1–100 GHz. Of great interest is an interpretation of the Nimbus-6 SMR data.

The Nimbus-6 SMR has been designed for continuous global mapping of the surface at the frequencies 22.235 and 31.400 GHz with a spatial resolution of about 150 km. Step-by-step scanning with SMR is done for six angles (on both sides of the sub-satellite trajectory) within the swath width ±53° with respect to nadir.

Theoretical modelling has shown that microwave emission of such media with small dielectric losses as ice and snow, is formed by a layer several metres thick and therefore depends substantially on the structure and characteristics of the upper layer (density, temperature, water content, salinity, size distribution of grains, seasonal stratification). Thus, satellite measurements of the T_B angular distribution at different frequencies and polarizations contain information which may be used to identify the types of snow and ice cover.

Fisher et al. (1976) considered the global-scale microwave images which illustrate possibilities of monitoring the seasonal evolution of the Earth's polar caps. The multi-year sea ice is easily identified from its lower emissivity and decrease with growing frequency ($\Delta T_B = T_{B,31.4} - T_{B,22.2} > 0$). As a rule, in the case of ice, $\Delta T_B < 0$ and only in winter are small positive values observed. T_B in the Antarctic is much below the surface temperature, since the emissivity varies here within the range 0.6–0.9.

The structure of the T_B field closely corresponds to the contours of snow accumulation. A zone of weak accumulation is clearly observed in the centre of Greenland. In the Antarctic, ΔT_B is characterized by a positive annual change in the summer, a decrease and transfer to small negative values in the winter.

The field of ΔT_B makes it possible to identify the firn against ice. The snow-covered land is characterized by a lower emissivity and negative ΔT_B. Fisher et al. (1976) suggested a theoretical model to simulate the transport of microwave emission in a scattering medium in the presence of 3-D random fluctuations of the refraction index as well as non-random variations of permeability, temperature and losses. The model is based on the combination of transfer theory and Maxwell equations in the Born approximation, which made it possible to calculate the external and internal T_B as a function of polarization and viewing direction for either semi-infinite or finite medium. The results of calculations show that an account of losses of sub-surface temperature profile determines a substantial change in the frequency-dependence of T_B.

The values of $\Delta T > 0$ in the Antarctic should be interpreted as determined by the fact that the size of snow and ice grains is greater than the "resonance" one at the frequency 22 GHz (0.8 mm). The annual change of ΔT_B with negative wintertime values is determined by decreased sub-surface temperatures and decreased losses in winter. The data obtained testify to possible retrievals of the thickness and water content of snow cover on land, as well as estimations of salinity and age of sea ice.

Electrically scanning microwave radiometers (ESMR) carried by Nimbus 5 and Nimbus-6 enable one to obtain surface images at 1.55 and 0.81 cm, respectively, with a spatial resolution of about 30 km. This made it possible, in particular, to monitor the spatial distribution and sea-current-determined dynamics of the antarctic sea ice. To test the GARP observation system, Zwally et al. (1976) drew weekly maps of sea ice, which characterized the distribution of open water (percentage) over the squares 2.5 by 2.5° lat. long.

Usually, individual ice fields and leads are below the ESMR resolution. However, it is possible to assess the ratio between the ice cover and open water within each element of resolution due to very different emissivities of ice and water. The observed T_B can be presented as a linear combination of the brightness temperatures of (εT_0) and water ($\varepsilon_w T_w$), with the contribution of atmospheric emission (A) taken into account:

$$T_B = (\varepsilon T_0)C + (\varepsilon_w T_w + A)(1 - C) \tag{6.1}$$

where C is the ice concentration. At the wavelength 1.55 cm, typical values are: $\varepsilon_w T_w = 120$ K, $A = 15$ K, and $\varepsilon T_0 = 250$ K.

The retrieval of the ice concentration from the data of measurements at one frequency is hindered by the variability of ε as a function of the type of ice surface temperature T_0, and by the variability of atmospheric microwave emission. In the zone of the Antarctic, it is possible to assess the ice cover concentration to an accuracy of $\pm 15\%$ and better.

The wintertime location of the southern boundary of sea ice in the Antarctic is controlled, first of all, by the antarctic circumpolar current, and not the prevailing wind field. The contours of this boundary clearly reflect the effect of sub-water ridges on the current. The existence of two stable zones of decreased concentration of the wintertime pack ice testifies to the presence of upwellings of warm waters ($T \approx 2°C$) to the surface. The ice concentration in the Ross Sea increases, for example, by 15–30%. But still more vivid manifestations of upwellings are a large extended polynya of about $0.25 \times 10^6 \, \text{km}^2$ near $0°$ E, which remained during the whole winters of 1974 and 1975. Here the ice concentration was below 15%. Since the heat- and moisture-exchange over the polynya is several orders of magnitude more intensive than over pack ice, the presence of this polynya should affect considerably the regional climate.

The use of active radar means permit one to obtain surface images with greater spatial resolution than in the case of passive radar. This has been confirmed, for example, by the Soviet–American "Bering Sea Experiment". Therefore, despite certain technical difficulties, SARs are being developed now to remotely sound the properties of natural formations. This refers also to the use of scatterometers.

Parashar et al. (1978) discussed the results of airborne testing of two scatterometers at frequencies 13.3 GHz and 400 MHz, carried out near Point Barrow at Alaska. The tests showed broad prospects for such an all-weather remote sounding technique to characterize the ice conditions and, in particular, to detect some types (thicknesses) of ice cover. Both scatterometers enable one to measure reflected signal within $\pm 60°$ along the aircraft route and they have the resolution in the transverse direction 3° and 7.5° for 13.3 GHz and 400 MHz of the scatterometers, respectively. Measurements were made on both orthogonal polarizations.

In accordance with the results of theoretical calculations, the data of measurements have shown that the multi-year ice (more than 180 cm thick) gives a stronger reflected signal at the frequency 13.3 GHz. The one-year thin ice (18–90 cm), on the contrary, is characterized by a much weaker reflection at this frequency. A very thin ice cover (less than 18 cm thick) exhibits a medium of reflection at the frequency 400 MHz. Water is identified at both frequencies.

An attempt to apply the data of scatterometers to identify seven types of ice (by age and thickness) has not given any positive result. Therefore the consideration has been confined to the following four categories: (i) open water; (ii) thin sea ice (5–18 cm); (iii) thick fresh and one-year ice (18–90 cm); (iv) thick one-year and multi-year ice (90–180 cm). In this case the problem of identification of the types of ice can be solved successfully, provided, the data of measurements on the vertical polarization for 12 viewing angles are used. The reliability of distinguishing between one-year and multi-year ice from the data for 13.3 GHz in May exceeds 90%; in April it constitutes 87%. The 400 MHz scatterometer provides the 75% reliability. With the viewing directions reduced to seven, the reliability reduces to 62%.

The SMMR on board the Nimbus-7 satellite acquired passive microwave snow data from 1978 to 1986. This instrument helped to improve understanding of the role of snow in global climate and in large-scale hydrology processes. Passive microwave data are sensitive to snow depth and water content. Retrieval algorithms use 18 and 37 GHz observations and assume snow density equal to 0.3 g cm^{-3} and snow-grain radius equal to 0.35 mm. Seven years of SMMR results indicate that the range in the four-month average (December–March, six-day intervals) northern hemisphere snow cover extent is near 5 million km^2 and the comparable range in average snow volume is near 60×10^{16} g.

Data collected by the SMMR of snow cover and snow depth for the northern hemisphere are presented in a map format—Figs 6.2 and 6.3 (see the colour section). The microwave maps are based on average brightness temperatures and displayed on colour-coded polar stereographic projection.

As seen from the microwave images the snow cover becomes stable first in northeast Siberia and in northern Alaska between mid-September and mid-October. From Figs 6.4 and 6.5 (see the colour section) we can see the annual variations of average northern hemisphere snow extent and snow volume over land (December–March).

The next stage of climate studies was connected with the use of DMSP/SSMI satellite observations data. Methodological problems of the improvement of the retrieval algorithms included the estimation of the effects of varying vegetation cover, stratigraphy of the snowpack, dielectric properties of snowpack under various stages of metamorphism and location.

6.4 RETRIEVAL OF SNOW AND ICE PARAMETERS: MODEL AND VALIDATION EXPERIMENTS

An important stage in summarizing the results of development of techniques for remote sensing of snow and ice cover was the COSPAR Symposium held in 1980 in Budapest.

To develop techniques for assessing the properties of snow cover from the data on its microwave emission, regular observations had been started in late winter of 1977 at the Swiss alpine station (the height above the sea level 2550 m), with the use of five radiometers operating at frequencies 4.9, 10.6, 21, 36, and 94 GHz and mounted on the 16-m tower. The antennas of all the radiometers provided the viewing angle 10°. Hofer and Good (1979) considered the result of measurements for six nearly four-day periods in May–June 1977, followed by a vast programme of *in situ* observations of the snow cover characteristics and meteorological parameters, including: the rate of accumulation and water equivalent of precipitation for the preceding 24 hours, total snow content, snow temperature at the surface and at a depth of 10 cm. Once in two weeks, profile measurements of temperature were made, as well as stratification, water equivalent, size distribution, crystalline structure and water content of snow at different depths.

To identify the properties of the upper 10-cm layer of snow, four classes of snow cover have been identified: (i) "fresh" snow; (ii) "old" snow; (iii) "ice" surface (snow, frozen upon intensive melting); (iv) "springtime" snow (eroded surface with holes and large round crystals more than 3 mm in diameter). For each of the four classes, three levels of humidity have been chosen: "dry" (without liquid water); "humid" (less than 3%

of liquid water), and "wet" (more than 3% of liquid water) snow. The data of measurements of the microwave emission of snow at two polarizations as a function of nadir angle, have been analysed with the use of correlation, factor and cluster analyses of vectors characterizing the totality of the data for different frequencies, polarizations and nadir angles 10° to 50° (in most cases, 20 vector components have been considered).

An analysis of the data revealed the possibility of identifying three types of snow: fresh or old "dry" snow, "dry ice" surface, as well as all (in total) types of "humid" and "wet" snow. Only in the case of "dry fresh" or "old" snow was observed a clear difference between T_B on horizontal and vertical polarizations. In other cases this difference was always less than the T_B variations caused by variations in the state of snow cover.

Sometimes a more detailed identification is possible, which determines good prospects for microwave remote sensing as a means of assessing the snow cover characteristics. The importance of microwave sounding is connected with its ability to determine the parameters, which cover the range of properties from those determined by the micro-crystalline structure to classic macro-scale characteristics. But this ability causes the main difficulty in the realization of the microwave technique, determined by the multi-parametric character of the snow cover microwave emission, which brings forth the necessity to further develop a theory describing the dependence of microwave emission on the complicated set of parameters.

In this connection, Shin and Kong (1982) noted that the prevailing contribution to the formation of T_B of natural formations was made by scattered microwave emission determined by inhomogeneous medium and the roughness of its interface (with a multi-layer medium). Two models of volume scattering by natural media have been suggested: (i) a model of a randomly inhomogeneous medium; (ii) a model of discrete scatterers immersed into a homogeneous medium. The interfaces have been considered here flat and, hence, the scattering on roughnesses has not been taken into account.

A theory has been suggested of emission transfer in the presence of a scattering layer located over the homogeneous semi-space limited by rough boundaries.

The choice of spherical scatterers made it possible to use the Mie formulas to calculate the phase functions and volume scattering coefficients. The roughness of interfaces is simulated by the introduced bistatic coefficients of scattering for the surface whose roughness is determined by the Gauss distribution, and the scattering on roughness is described as that obtained by combining the Kirchhoff approximations and geometric optics.

The introduction of bistatic scattering coefficients makes ambiguous the calculations of the emissivity of the medium. Therefore calculations were made of the lower and upper limits of emissivity based on the use of two alternative formulations.

The upper limit of emissivity is obtained through calculations of bistatic scattering coefficients for a layer limited by rough surfaces, with the subsequent integration by the scattering angles within the upper hemisphere and the use of the invariance principle, whereas the lower limit of emissivity is estimated through direct calculations of microwave emission on the assumption of the temperature homogeneity of the medium.

The equations of radiation transfer have been integrated with the use of the Gauss technique, and the results obtained have been presented as T_B dependences on viewing angles for the orthogonal polarizations. An analysis of calculations results has shown that the presence of a rough lower boundary leads to an increase of T_B, except for large zenith viewing angles for the vertical polarization. The roughness of the interface upper boundary determines the "flattening" of the angular radiance distribution and a decrease of the difference of radiances on horizontal and vertical polarizations.

There are two principal circumstances that hinder the use of the remote sensing technique considered: (i) classic parameters of snow cover and techniques for their estimation are inadequate from the viewpoint of simulation of the dependence of microwave emission on the most essential parameters; measurements are needed of the vertical profiles of temperature, humidity, mass share of moisture and crystalline structure with a vertical resolution exceeding the applied wavelength, which is practically impossible; (ii) there is still no complete theoretical model of microwave emission of snow.

Based on existing experience, a three-channel system of microwave sounding with a high-frequency channel in the window 30 GHz and radiometric frequencies, differing by a factor of 3, can be considered optimal. Measurements should be made on both polarizations, but it is sufficient to make them only for two nadir angles. In further studies, the emphasis should be placed on the data about the depth of penetration of microwave emission through the snow and on the further search for various versions of remote sounding of snow cover.

Having analysed the data of observations, Mätzler et al. (1982) drew the conclusion that possible identification of three seasonal types of snow. The wintertime snow cover is characterized by the absence of the melting-induced metamorphism. At the frequency 36 GHz, a clear dependence is seen of T_B on the water equivalent of snow, with strongly decreasing T_B in the case of thin snow cover, but with linearly increasing T_B, when the water equivalent grows for the range of thicknesses 20–50 cm. The T_B decreases with growing frequency at frequencies exceeding 10 GHz. In the range of smaller frequencies the soil contributes substantially to the formation of microwave emission.

The springtime snow is characterized by the effect of the diurnal cycle of melting and freezing. The snow cover is called the springtime one when its upper layer several centimetres thick suffers a considerable metamorphism due to melting at the expense of either air temperature increase of daytime insolation and nocturnal freezing. The phases of melting and freezing of the snow surface layer are easily to identify due to total inversion of the respective spectral dependence of T_B near the wavelength 2 cm (especially on horizontal polarization which is explained by the film of liquid water formed during the melting), almost opaque for microwave emission. The prevailing effect of scattering near the surface determines in this case a decrease of T_B on horizontal polarization with decreasing frequency and growing viewing angle. The frozen snow is characterized by T_B decreasing with growing frequency, manifestly stronger than in the case of the wintertime snow.

In the summer the whole snow cover is saturated with water and this determines high values of T_B, especially at frequencies exceeding 10 GHz. The microwave emission spectrum is similar to that observed in the case of the melting springtime snow and is characterized by a substantial dependence on the water equivalent and by the absence of diurnal change. With precipitation, the surface roughness grows, which weakens the T_B contrast at different polarizations. The use of the dependence of T_B on the water content of snow permits the monitoring of the dynamics of rain infiltration into the snow cover layer.

The results obtained suggest the conclusion with respect to requirements for the data of microwave measurements which make it possible to determine various characteristics of snow cover: (i) to distinguish between dry snow (in winter or frozen springtime snow) and moist snow and to estimate the water equivalent in the period of the wintertime snow accumulation, data are needed of measurements at not less than two frequencies, in the range 10–40 GHz; (ii) the existence of data for a channel in the frequency range less than 5 GHz is important from the viewpoint of assessing and revealing the hydrological conditions; (iii) of great importance is the monitoring of the T_B diurnal change; (iv) the T_B values on horizontal polarization in the range of viewing angles 30–50° are most sensitive to hydrological parameters.

Ulaby and Stiles (1980) performed ground-based observations using a microwave active spectrometer (MAS) to study the possibilities of interpreting the data of scatterometric measurements for the remote sensing of the snow cover characteristics.

The MAS observations made at different locations in the USA enabled one to obtain a vast database of radar measurements of the backscattering coefficient, σ°, at frequencies in the range 1–18 GHz, as well as at 35.6 MHz. At the same time, microwave measurements of T_B were made at 10.7, 37, and 94 GHz. Also, control measurements were made of different parameters of snow cover: thickness and stratification, vertical profiles of density and temperature, humidity (relative volume content of liquid water per unit snow volume) and water equivalent (thickness of the equivalent layer of liquid water). The soil temperature was recorded at various depths, and in the periods of soil thawing, its water content was estimated. Atmospheric pressure, air temperature and humidity were continuously registered.

The main objective of the observation programme was to study the dependence of radar backscattering coefficient, σ°, and emissivity, ε, on the thickness of snow cover, humidity and water equivalent, surface roughness and some other parameters. The data of T_B measurements showed that with the humidity of the upper 5-cm snow layer increasing from 0% to 2%, the T_B slightly rises at a frequency of 10.7 GHz, whereas the T_B at 37 GHz grows more than 100 K.

With the snow cover 26 cm thick (the water equivalent 6.2 cm) the backscattering coefficient at 2.6 GHz is mainly determined by the soil, but with the frequency increasing to 35.6 GHz, the effect of snow moisture shows itself more markedly. The emissivity for dry snow decreases with frequency (this is due to the growing effect of scattering), and in the case of moist snow, only a weak dependence of frequency is observed.

The results of observations shows that the σ° increases and the ε decreases with an increasing thickness of the dry snow cover (water equivalent) until the vertical extent of snow cover becomes equivalent (from the viewpoint of propagation of electromagnetic emission) to a semi-infinite medium. The presence of liquid water in snow leads to further attenuation and decrease of scattering, which determines the decrease of σ° and increase of ε. The use of spectral dependence of σ° and ε enables one to practise the remote sounding of the humidity and water equivalent of snow, especially in the case when the data are available from both daytime and nocturnal observations.

An analysis of requirements for the sea ice information has led to the conclusion that the data are needed both with high resolution of the order of 100 m for spatial scales of about 100 km (steering of ships and other application problems) and with the 10 km resolution for sub-global and global scales (climate studies). In the case of snow cover the required resolution and spatial coverage vary from 100 m and tens of kilometres (hydrological processes over the catchment area) to several kilometres and large scales, up to global scale (climate). The most important parameters (apart from the ice and snow cover extent) are as follows: the type (including age) and concentration of ice cover, the state (e.g. melting) and water equivalent of snow cover. Major characteristics of glaciers are their extent, depth, topography and the rate of accumulation.

Special emphasis has been placed (within the MIZEX programme mentioned above) on the use of techniques of passive and active microwave sounding near the ice cover edge, where its dynamics is very intensive (Johannessen, 1987; Johannessen et al., 1992). It is particularly important that these areas are developed to enable studies of coastal processes (Johannessen et al., 1994, 1995).

Continuous satellite microwave observations as well as processing the visible and IR data (including information from Landsat and SPOT satellites) have made it possible to accomplish a regular global mapping of sea ice and continental snow cover, to draw maps of sea ice concentration, as well as to study a detailed regional dynamics of the cryo-sphere. For the 1981–87 period the microwave data do not reveal any marked global trend of sea ice extent. The prospect for further development of remote sounding of ice and snow cover are connected with the use of space-borne radars, which will make it possible to obtain data with a resolution of tens of metres and better. In this connection of great importance are the results of the Soviet studies, especially those obtained with the synthetic aperture radars carried by the satellite "Kosmos-1500" and "Kosmos-1870", "Almaz", and "Almaz-1" as well as the ERS-1 and JERS results.

Extensive information on the cryospheric dynamics obtained from satellite data for the period 1979–86 enables one to accumulate the global database on the concentration of sea ice with the resolution of 50 km from observations with the Nimbus-7 SMMR. The DMSP satellites transmit continuously the visible images with a resolution of 0.6 km. A system of arctic sea buoys supplements this information with the data on the ice drift and atmospheric pressure near the surface.

The DMSP data were used in the regular (every three days) mapping the ice cover in the summertime Arctic. Average surface albedos for the Arctic were obtained—0.75 (May), 0.63 (June), 0.46 (July), and 0.41 (August). From mid-August, the albedo starts increasing in the central arctic water basin due to snowfalls, but in the adjacent seas it can decrease (a minimum of the sea ice extent is usually observed in September). There is a substantial interannual variability of albedo (respectively, absorbed solar radiation), which can explain the melting-induced variations in the thickness of ice cover within 0.66 m.

An analysis of microwave data for the period 1979–84 for the Canadian water basin revealed in some regions a decrease of ice concentration, reaching 70–80%, caused by the effect of quasi-stationary cyclones, which brings forth a divergence of sea ice concentration by about 0.5% per day. The appearance of the open water areas should cause a considerable intensification of the sensible heat exchange between the atmosphere and the ocean, which should be taken into account in the numerical modelling of the climatic impact of the greenhouse effect and other factors.

An analysis of the SMMR data for nine seasons revealed differences of the annual change of sea ice in the northern and southern hemispheres caused, mainly, by inhomogeneous distribution of land and ocean. The most peculiar difference is an asymmetry of the annual changes manifested through a faster decrease of the ice cover extent in the summertime southern hemisphere due to a very unstable thermocline in the southern hemisphere.

An examination of the large-scale cryospheric dynamics in five sectors of the Arctic from the data for 15 years revealed the existence of a considerable interannual variability (especially in the Antarctic) and asynchronous variations in different regions. During nine years (1977–87) neither in the Arctic nor in the Antarctic was there observed a marked secular trend of variability of the annual change of the ice cover extent, but the global extent decreased by about 5% due to changing phase and shape of seasonal oscillations. From the data of Wadhams (1990), based on the use of results of the sub-ice acoustic sounding (from a submarine) in May 1987 and October 1976 along the route more than 400 km north of Greenland, the thickness of ice cover decreased, which was equivalent to a 15% loss of ice volume over the area 300 000 km^2. Note should be taken, however, that the interpretation of data on cryospheric dynamics from the point of view of revealing climatic variations should be very cautious.

The processing of a ten-year (1978–87) global SMMR data series made by Gloersen and Campbell (1991) made it possible to characterize the statistical features of the global sea ice cover dynamics. Fourier analysis revealed not only the prevailing peak corresponding to the annual change, but also considerable second and third harmonics as well as marked fourth and fifth harmonics. The filtering out of periodic variability has led to the conclusion about the existence of a negative trend of the extent of the arctic ice cover for nine years equal to $1.9 \pm 1.3\%$ with a 90% statistical significance at a level of $2s$ (the ice cover extent was assessed in processing the satellite microwave data as a surface area outlined by the isoline of the ice cover concentration equal to 15%).

In the case of the antarctic ice the trend is absent and its global mean value is $1.0 \pm 0.7\%$. The area of the open water surface in the Arctic varied within the range (1.6–3.2) $\times 10^6$ km^2, constituting about 10% in the period of maximum ice cover extent. The respective values for the Antarctic are (1.5–4.5) $\times 10^6$ km^2 and 25%.

The operation of the Geosat satellite has opened up possibilities of monitoring the topography of ice sheets and sea ice from the data of radioaltimetric observations (Lingle et al., International Conference, 1991): for example, the annual change of the surface level in the zone ablation in west Greenland. To reveal the climatic trends, the scientists of the US Geological Service started in 1978 a regular processing of the Landsat data to monitor the dynamics of glaciers in different regions of the globe (Williams, International Conference, 1991). Molina (International Conference, 1991) emphasized, however, that the retreat of glaciers cannot always serve as an indicator of climate warming, which determines the necessity of a thorough analysis of the data of such observations.

The results of monitoring the ice cover dynamics from the DMSP SSMI data at frequencies 18, 37, and 85 GHz made it possible to obtain the ice cover concentration, the share of multi-year ice, and the location of ice edge in both polar regions. A joint processing of the satellite microwave information and the data of drifting buoys has opened up possibilities of study of the interaction between the atmosphere and the ice cover on synoptic scales in the Arctic. The ice cover microwave data interpretation becomes more informative with the use of polarization radio and gradient ratio at frequencies 18 and 37 GHz.

The combined use of the SMMR and AVHRR data (for 8 years) has provided a reliable retrieval of the ice condition characteristics. Owing to the difference between PR and GR for one- and multi-year ice, it is possible to determine the age of ice. In the future, the physical algorithms of retrieval are planned to be used.

Jezek and Cavalieri (Role..., 1990) showed that the T_B difference on vertical polarization as well as horizontal can be an indicator of the presence of free water in the ice firn.

Naturally, the requirements for the observational data in the Arctic are not confined to information on the ice and snow cover or permafrost. The basic parameters of interest are as follows: temperature and heat balance of the ocean surface, chlorophyll (phytoplankton) concentration, the area of the open water surface along the arctic coastline, geographic distribution of ice cover concentration and thickness, structure and extent of glaciers, spatial and temporal variability of snow cover (including information about its thickness, density and water equivalent), temperature and radiation budget of land surface, extent of permafrost and zones of its instability, soil moisture, atmospheric pollutants, vertical profiles of radiative characteristics of the atmosphere, amount and type of clouds, precipitation, stratospheric temperature, content of ozone and other minor gaseous components.

Informative materials on these problems were presented at the IUGG Assembly in Vienna (IUGG..., 1991). A great interest in the climate-forming role of polar regions determined the inclusion in the programme of the Assembly of some symposia dedicated to both these problems and to specific features of high-latitude remote sounding.

In his informative overview, "Remote sounding of snow and ice: overview and present state", Carsey (1992) summarized the intensive development during the last decade of the techniques of remote sounding of snow and ice cover (sea ice, glaciers), with emphasis on the analysis of the effect of microphysical characteristics of snow and ice on the formation of the radiation field in different wavelength regions as well as on the justification of requirements for the data of the remote sounding of snow and ice cover.

In the overview paper, "Polar sea ice variability from spaceborne observations", Cavalieri gave interesting data on the latest observations of the spatial and temporal variability of the ice cover extent. He emphasized, for example, that in the annual change the northern hemisphere ice cover extent varied from a minimum of $8 \times 10^6 \, \text{km}^2$ in September to a maximum of $15 \times 10^6 \, \text{km}^2$ in March. In the southern hemisphere the ice cover reached 54° S, and its extent varied from $4 \times 10^6 \, \text{km}^2$ (February) to $10 \times 10^6 \, \text{km}^2$ (September).

In their contribution on ice cover microwave remote sensing Weaver et al., summarized the results of monitoring the ice cover dynamics from the DMSP SSM/I data at frequencies 18.37 and 85 GHz. The results of T_B measurements at these frequencies were used to retrieve the ice cover concentration, the share of multi-year ice and the location of ice edges in both polar regions.

7

Remote sensing of ice and snow cover: the Russian experience

7.1 MULTI-SPECTRAL MICROWAVE REMOTE SENSING OF FLOATING AND CONTINENTAL ICE COVER, SNOW, AND FROZEN SOILS: PRINCIPAL DIRECTIONS OF DEVELOPMENT

The all-weather character of microwave remote sensing of the surface parameters and the possibility of the quantitative interpretation of the results obtained have brought forth the development of techniques to interpret the microwave images. Studies of the state of various types of surface in the microwave region were started in the Soviet Union in the mid-60s. Encouraging results were discovered after the analysis of first "Kosmos-243" microwave measurements (Basharinov et al., 1974). Soon after the launches of several Soviet microwave satellites, the programme was worked out of multi-year theoretical and observational studies aimed at the development of satellite techniques and the means to solve various problems of meteorology, oceanography and land use. The National Inter-disciplinary Programme has implied the financing of the existing and newly formed scientific groups, operating in the systems of Hydrometeorological Service, Academy of Sciences, and some other governmental departments.

The scientists of the Main Geophysical Observatory (MGO) have been the first to develop microwave techniques of the remote sensing of sea ice (Kondratyev et al., 1973b, 1992b; Rabinovich et al., 1970; 1975). The development has been first of all aimed at creating a complex of airborne instruments to map the state of the drifting ice cover. First flights of the Ilyushin-18 aircraft allowed one to discovered relationships between the T_B of the system "sea surface–atmosphere" at $\lambda = 3.2$ cm and the concentration of sea ice independent of its age (Fig. 7.1) (Rabinovich et al., 1970) as well as to map the distribution of ice and its properties along the flight route of the aircraft.

The instrument operated using a parabolic antenna with a mechanical scanning mirror. The in-flight information was film-registered. The in-flight film developing enabled one to map the ice concentration in the region of flight. The problem has been discussed of developing an instrumentation that will transmit the microwave data to ships sailing in the Arctic region. Unfortunately, the plans to put the microwave survey into the practice

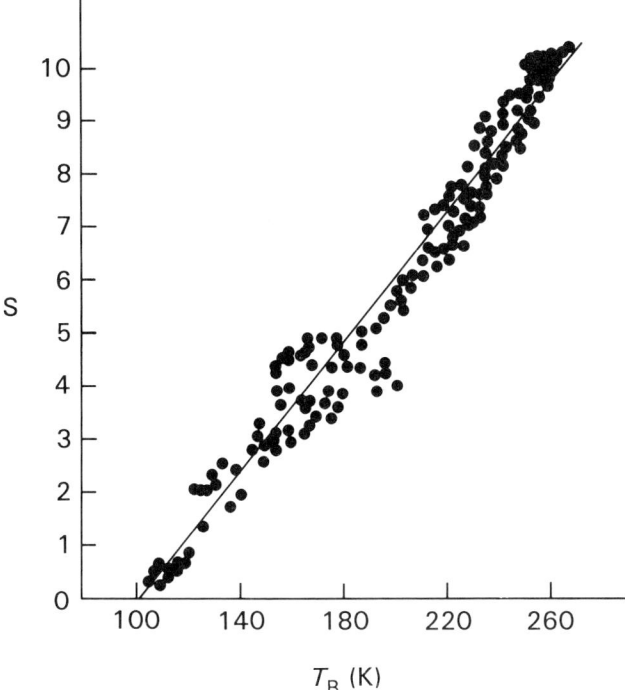

Fig. 7.1. Brightness temperature versus sea ice concentration (different age of sea ice)
(Rabinovich et al., 1970)

of ice navigation have not yet been realized: the microwave instruments were too sophisticated, and ships' technical staff were not sufficiently well-qualified.

In this connection, attempts have been made to advance the instruments in order to
create an airborne microwave complex to obtain radioimages of the ice cover, to perform
in-flight operational image processing, decoding, and preparing of ice maps.

Model calculations of the sea ice emittance with the use of the data on the dielectric
constants of different structures of ice made it possible to substantiate the choice of
conditions of radiomapping the arctic seas. The data in Figure 5.6 suggest the conclusion
about the use of vertical polarization in the survey for all types of the ice at any thermodynamic temperature, the change of emittance $\Delta\Sigma_V/\Delta\Theta$ is small and below the absolute
value of $\Delta\Sigma_H/\Delta\Theta$. However, as the results of the Ilyushin-18 observations have shown,
with the survey conditions chosen in this way, the brightness of the edges of the
radioimage can rise. The analysis of the airborne information has shown that on the
horizontal polarization a decrease of brightness towards the frame edges is observed only
for some types of ice, that is, it is possible to develop an operational technique to study
the arctic seas without compensation for the angle dependence of ice cover emissivity.

The scanning radiometer made it possible to obtain the spatial distribution of sea ice,
to identify the open-water surfaces, to assess the size of ice floes and leads, to estimate
the ice concentration, and to outline the water–land interface in conditions of

cloudiness—these parameters are very important for arctic climate study. The processing of observational data for radiomapping the ice concentration in unfavourable meteorological conditions—cloudiness, precipitation, haze—was carried out on the Kara Sea and Laptev Sea in the period of navigation.

As is seen from the microwave survey, in the region of Sverdrup Island three systems of sea ice have been decoded—Fig. 7.2. The first system of ice cover on this map is the fast ice adjacent to the island. The minimum extent of the near-surface ice constituted 4.5 km; this area was adjacent to the south-eastern point of the island.

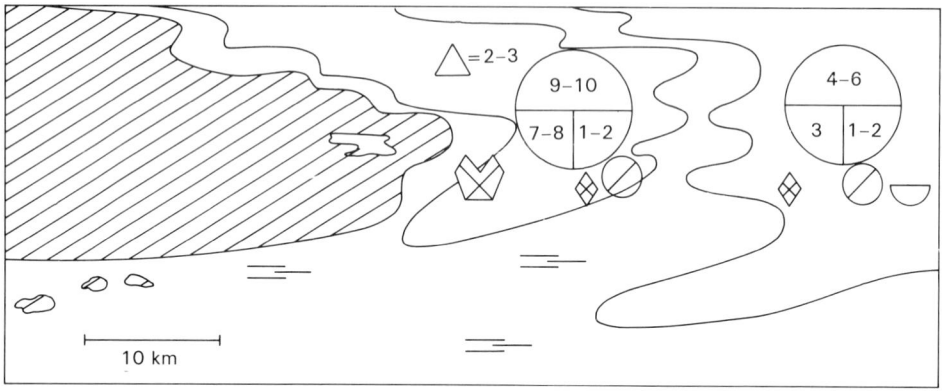

Fig. 7.2. Map of sea ice concentration derived from $\lambda = 3.2$ cm microwave survey of the Kara Sea ice cover (Kondratyev et al., 1992b)

The second system is a field of large-floe pack ice 120–140 cm thick, 10–12 km in length, like a half-broken ring, surrounding Sverdrup Island, clinging close to the fast-ice zone in its southwestern part. The minimum width of the identified circular leads, adjacent to the area of the minimum near-shore ice extent, constituted here 200 m, which was sufficient for navigation purposes. Ice concentration in this zone is 70–80%.

The third circular system is the destroyed first-year ice: fields of large-broken ice, surrounded by smaller ice floes; the diameters of the identified ice floes vary from tens to hundreds of metres. Ice concentration in this zone is 40–50%. Note that a microwave image of Sverdrup Island enables one to monitor the genesis of pack ice in the observed region—the configuration of rings, outlines of ice floes and clusters of the fields of large and small ice floes, which suggests the conclusion that this ice has formed in this region and in the previous winter period it had been adjacent to the island. Unfortunately, the microwave survey data contained inadequate information about the state and the thickness of sea ice, which can be judged only indirectly.

On the basis of analysis of this microwave image, the following structures can be identified: the open water surface, homogeneous unbroken fast-ice zones, and island surface differing in emissivity from the adjacent ice area.

Another level of development of the microwave ice mapping was to develop a system of radiometric instruments with an conical electric scanning antenna—phase grid (Alexandrov et al., 1986).

The Institute of Radio Engineering and Electronics of the Academy of Sciences (Moscow) carried out a number of theoretical and observational studies of the emittance properties of sea ice of different thicknesses and age, which proved the usefulness of microwave remote sensing (Basharinov and Gurvich, 1970).

Of interest is the programme of studies of the electrophysical parameters of sea and continental ice, snow and frozen soils, as well as thematic processing of satellite and airborne measurements, accomplished in the Arctic and Antarctic Research Institute (Bogorodsky and Oganesian, 1987).

The associated developments should be mentioned in creating the techniques to study multi-year ice carried out in the National Center on Studies of Natural Resources of the Russian Committee on Hydrometeorology (Moskow) (Bukharov et al., 1990).

Theoretical and observational studies of the parameters of the state of the ice and snow cover of various types of inland water basins have been made in the Institute for Lake Research of the Russian Academy of Sciences in St. Petersburg (Kondratyev et al., 1985, 1989).

7.2 "BERING SEA EXPERIMENT (BESEX)": SUB-PROGRAMME OF THE SEA ICE PARAMETERS STUDY

Within the American–Soviet cooperation in space, a joint USA/USSR expedition was organized in 1973, aimed at the solution of problems of microwave remote diagnostics of the polar ocean and atmosphere parameters (Kondratyev et al., 1973a; USSR/USA Bering Sea Experiment, 1982). Synchronous meteorological, oceanological, radiation, and microwave measurements were made to evaluate the techniques of identification of precipitation zones, ice cover characteristics, determination of temperature and roughmess of the sea surface.

Multi-channel microwave measurements of thermal emission at different frequencies were performed from the aircraft synchronous with meteorological, aerological and oceanological observations from ships and satellites to obtain a vast volume of both *in situ* and remote sounding data for intercomparison and interpretation of microwave information. During the USSR/USA experiment the following basic problems were solved:

— microwave measurements have been made of the open sea surface at various temperatures and for different sea states, of the ice cover, and of the liquid precipitation zones in centimetre and millimetre lengths with the help of the equipment carried by Russian and American aircraft;

— ship measurements have been made of the thickness, structure, age, and dielectric parameters of sea ice, the temperature and the roughness of sea surface, the water content of the atmosphere; complete information about the physical state of the atmosphere and cloudness has been obtained by means of aerological (ship) and radiation (aircraft) sounding;

— procedures of microwave measurements and diagnostics of parameters of the ice cover, the atmosphere and the sea surface have been intercompared, the accuracy of determination of meteoparameters obtained from microwave measurements and direct measurement data has been evaluated.

Based on the climatological analysis of the Bering Sea region, the site of the experiment was chosen in the international waters of the Bering Sea equally removed from the USA and the USSR. The expedition lasted from 15 February to 7 March 1973. This choice made it possible to accomplish the whole programme of investigation of the important arctic climatic parameters mentioned above.

Fig. 7.3 demonstrates the location scheme of the experiment area. The USSR used Cape Schmidt as the base for its aircraft; the USA used the airport at Anchorage for this purpose. To obtain the *in situ* observational data, the USSR used a weather-ship "Priboy" and the USA used the icebreaker "Staten Island". In this case the USA scientists could perform the ice observations in the northern part of this area, and the Russian scientists at the ice edge and in the ice-free part of this area—Fig. 7.4.

The flying laboratories performed microwave measurements—the Ilyushin-18 of the Voeikov's Main Geophysical Observatory and the Convair-990 of the Goddard Space Flight Centre, NASA.

The ice cover study sub-programme connected with obtaining the microwave and infrared images of the ice area requires each aircraft to fly a pattern in its respective grid area over the ice region. Fig. 7.5 demonstrates the 5 March 1973 flight tracks of both aircraft. Each grid area was 100×100 km in size, and the two grids have a 30×30 km overlap at one corner. At the point of "Priboy's" location the ice was of different age and solidity, but not exceeding a concentration of 40–50%.

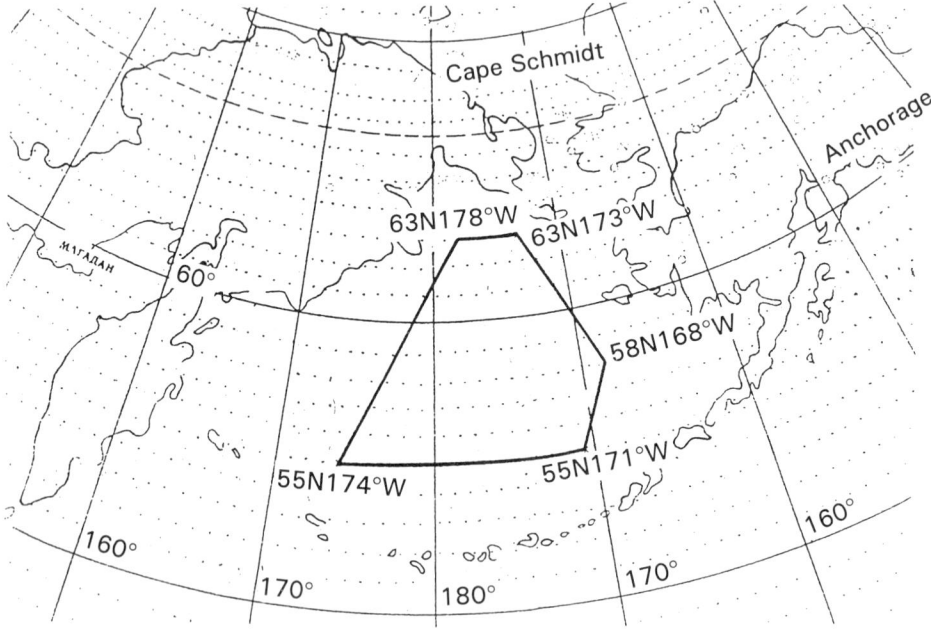

Fig. 7.3. Location scheme of the Bering Sea Experiment (BESEX) area (Kondratyev et al., 1973a).

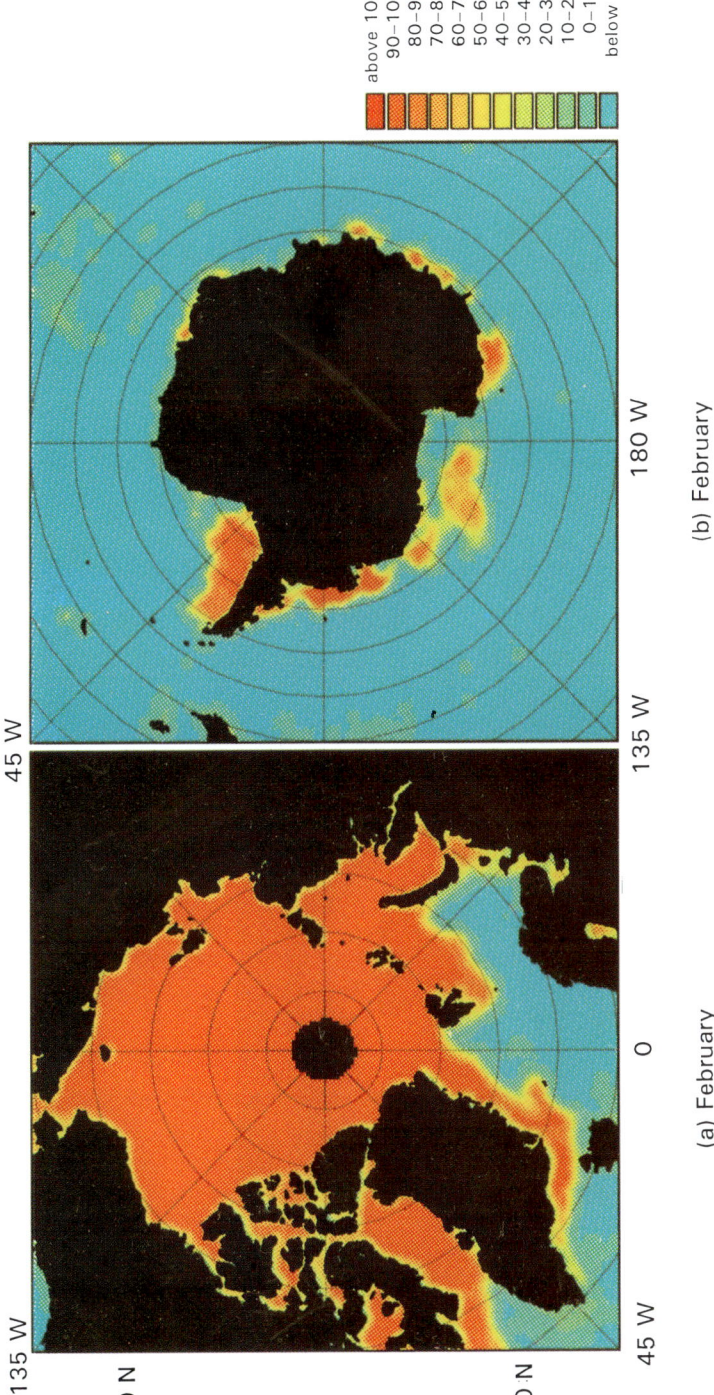

above 100
90–100
80–90
70–80
60–70
50–60
40–50
30–40
20–30
10–20
0–10
below 0

45 W

(b) February

180 W   135 W

(a) February

45 W   0

135 W

60 N   60 N

Fig. 6.1. Total sea ice concentration, northern and southern hemisphere: (a), (b) February; (c), (d) August (Parkinson and Gloersen, 1993)

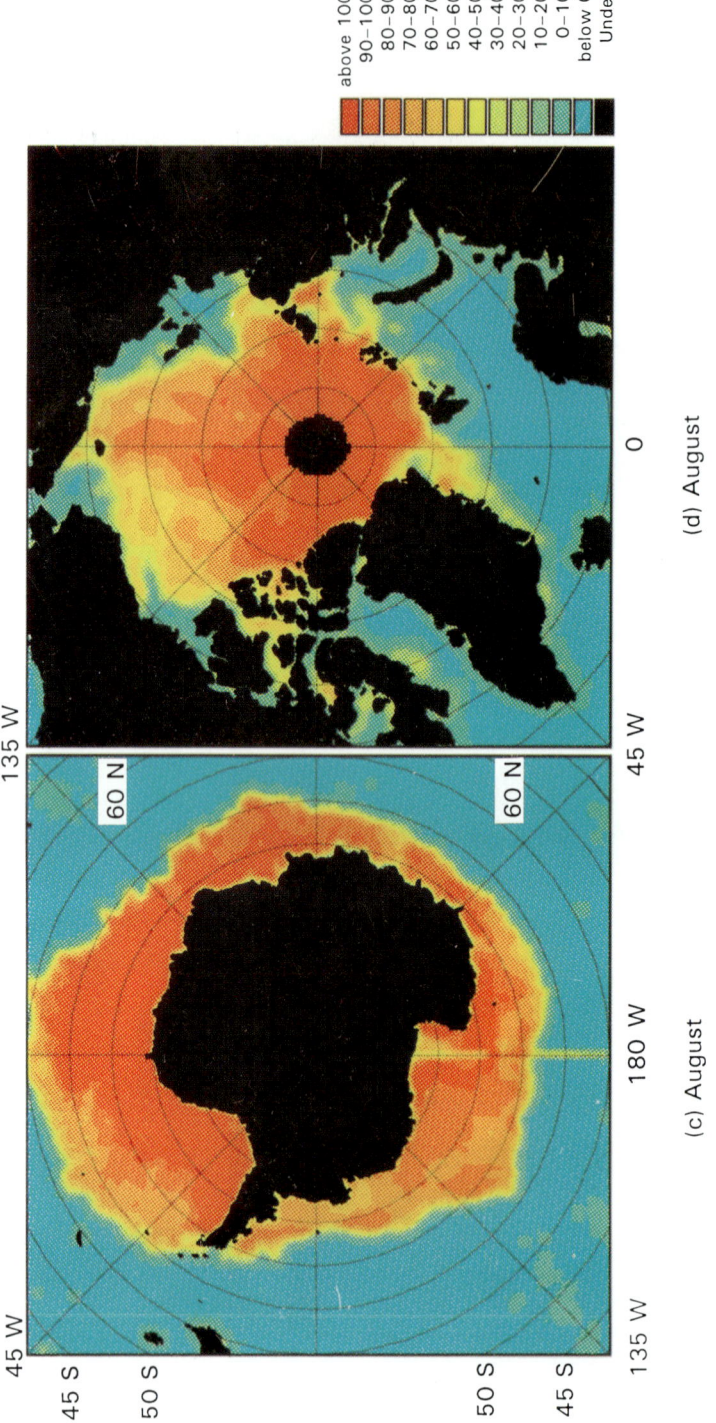

above 100
90–100
80–90
70–80
60–70
50–60
40–50
30–40
20–30
10–20
0–10
below 0
Undef

(c) August

(d) August

Fig. 6.1. Total sea ice concentration, northern and southern hemisphere: (a), (b) February; (c), (d) August (Parkinson and Gloersen, 1993)

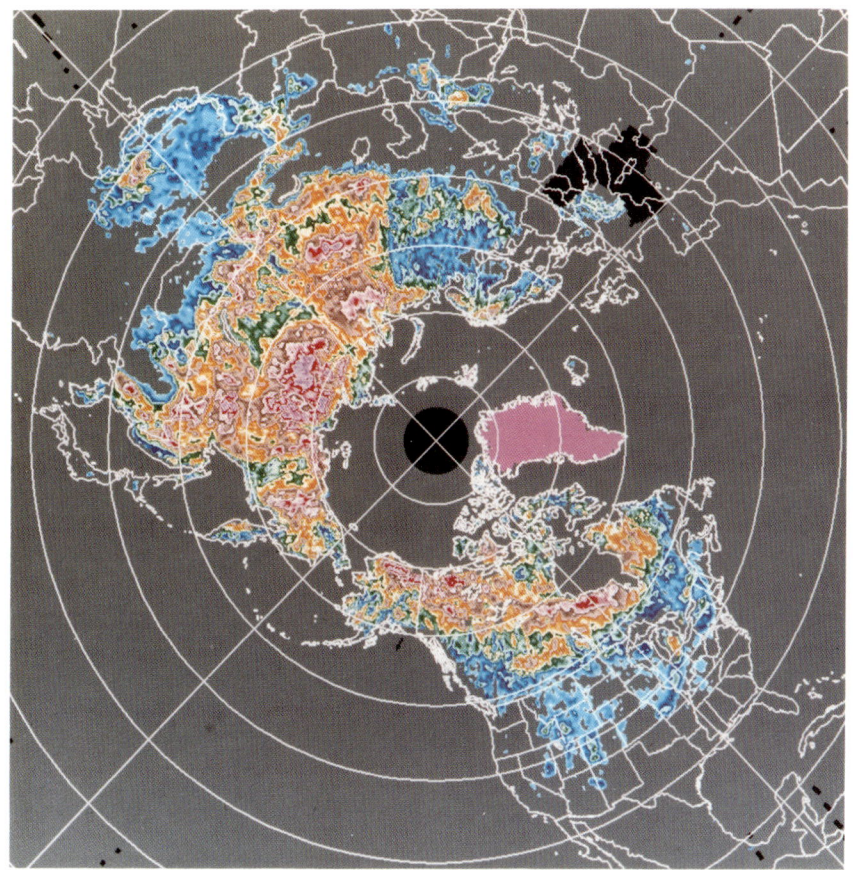

SNOW DEPTH $1.59 \times (T_{18H} - T_{37H})$ cm

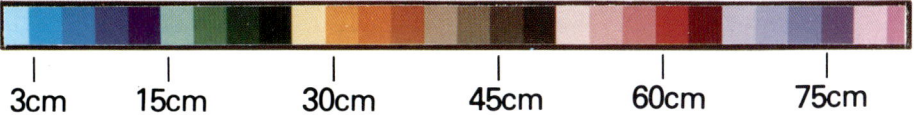

| 3cm | 15cm | 30cm | 45cm | 60cm | 75cm |

Fig. 6.2. Nimbus-7 SMMR derived snow depth map (February), northern hemisphere.
February, 1981 (Foster and Chang, 1993)

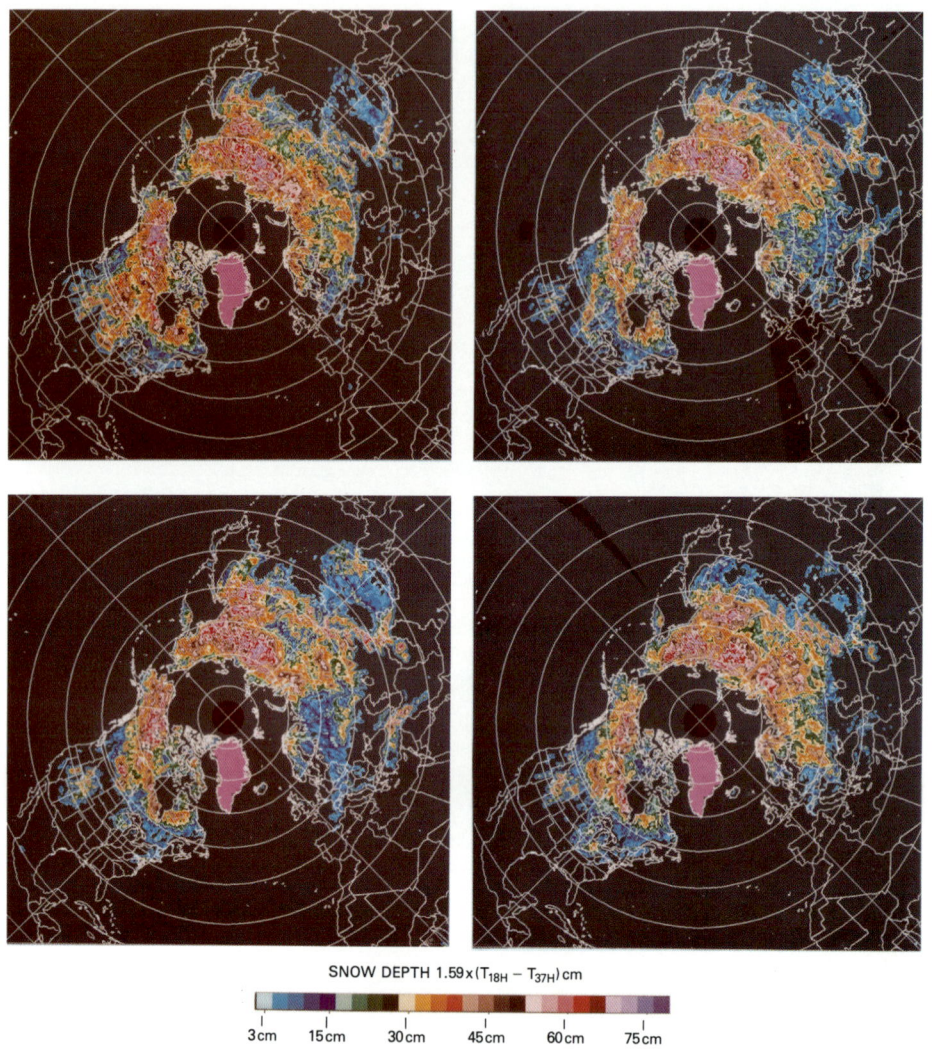

SNOW DEPTH $1.59 \times (T_{18H} - T_{37H})$ cm

3 cm　　15 cm　　30 cm　　45 cm　　60 cm　　75 cm

Fig. 6.3. Nimbus-7 SMMR derived snow depth maps (February): (a) 1979; (b) 1982; (c) 1983; (d) 1985 (Foster and Chang, 1993)

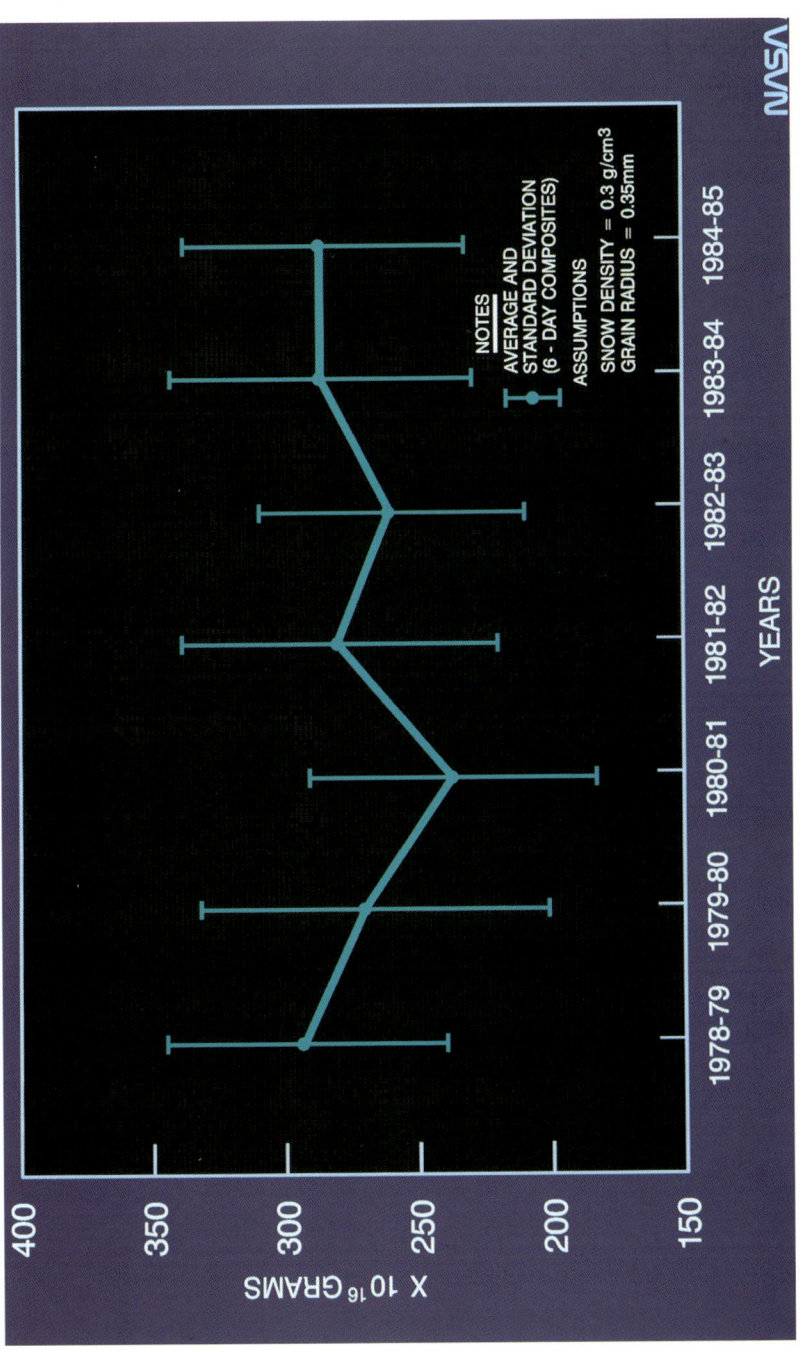

Fig. 6.4. Average northern hemisphere snow extent over land (December–March) (Foster and Chang, 1993)

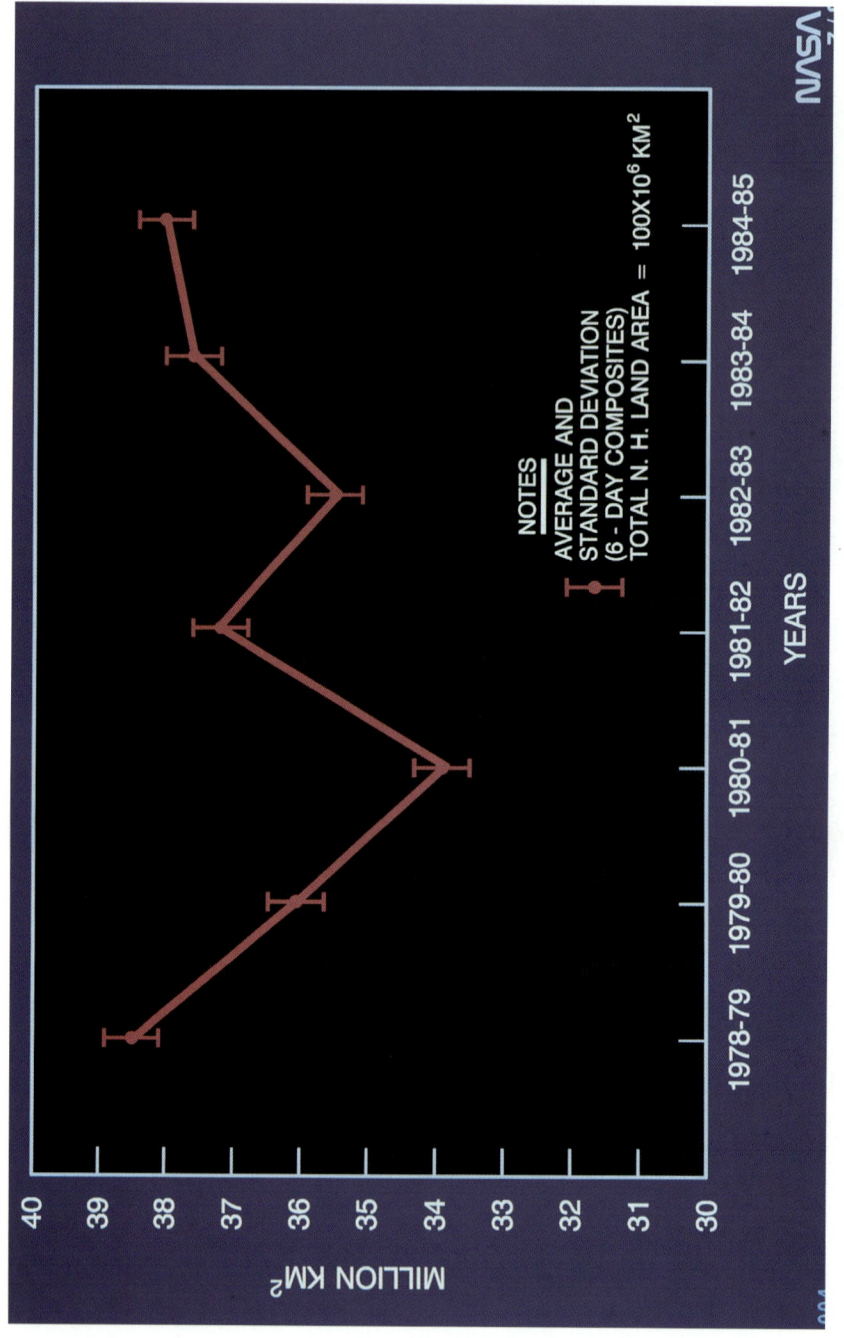

Fig. 6.5. Average northern hemisphere snow volume over land (December–March) (Foster and Chang, 1993)

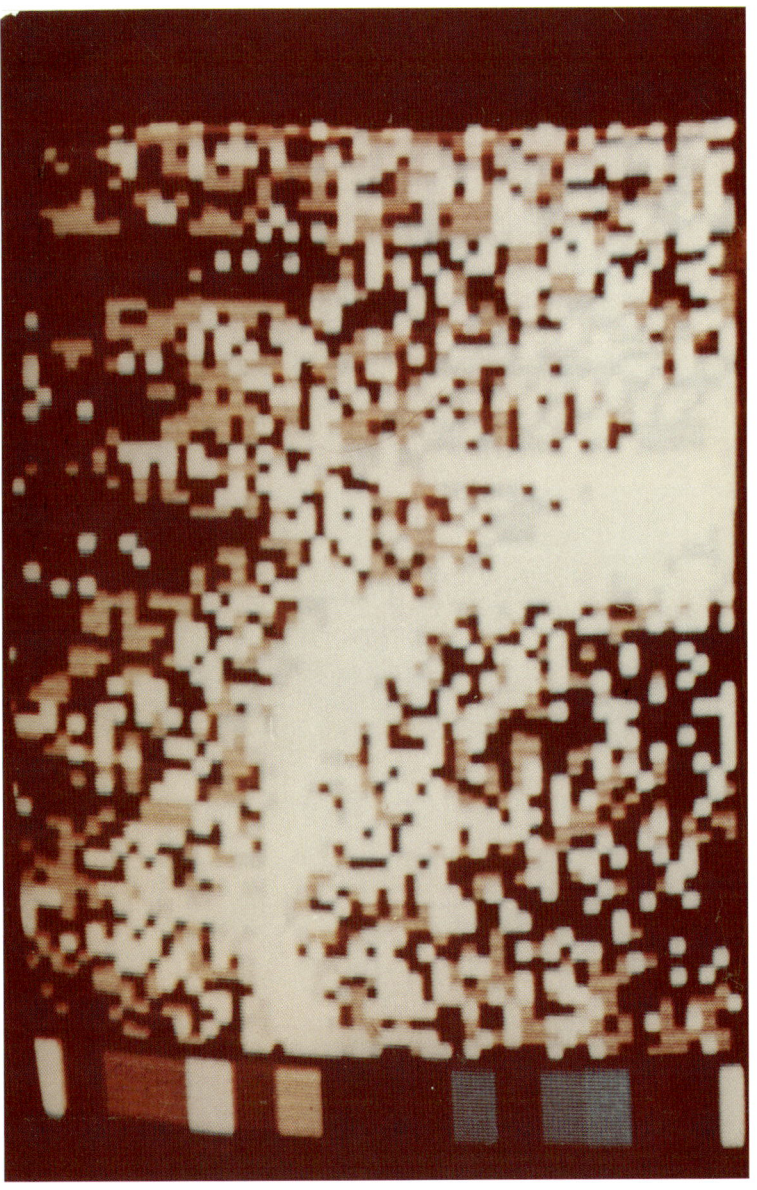

Fig. 7.6. Microwave image of consolidated pack ice ($\lambda = 3.2$ cm) (Kondratyev et al., 1992b)

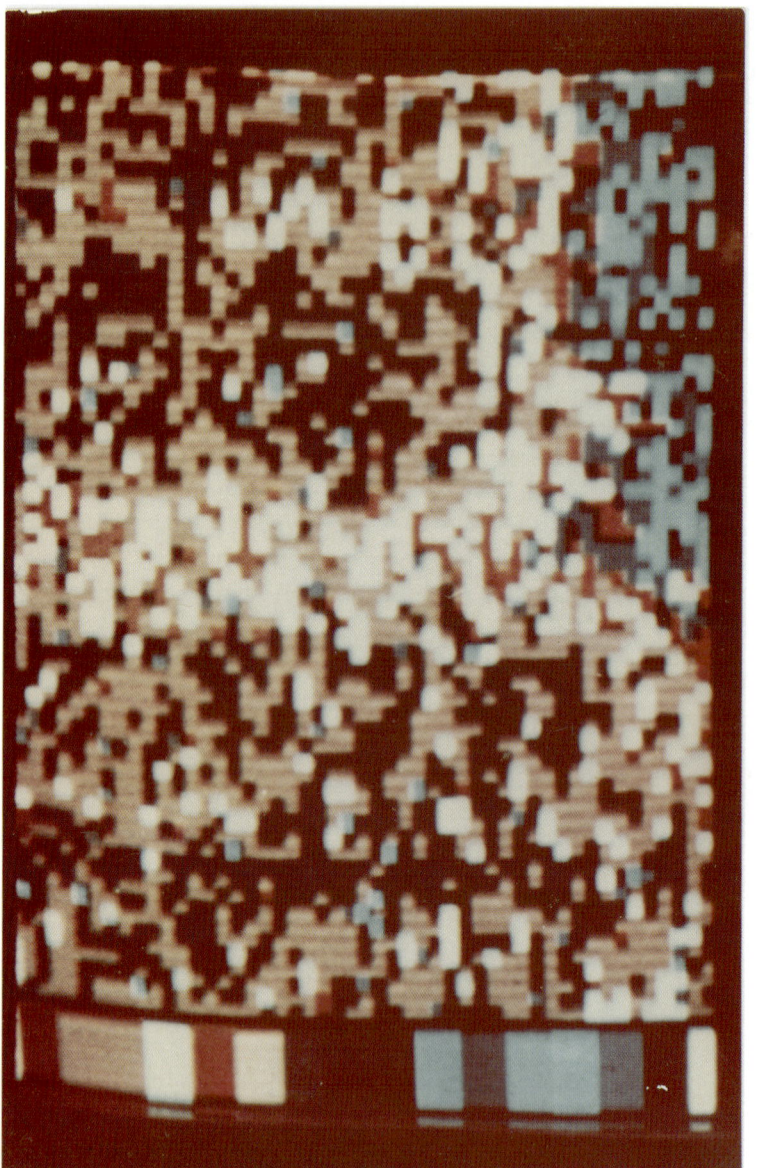

Fig. 7.7. Microwave image of ice edge ($\lambda = 3.2$ cm) (Kondratyev et al., 1992b)

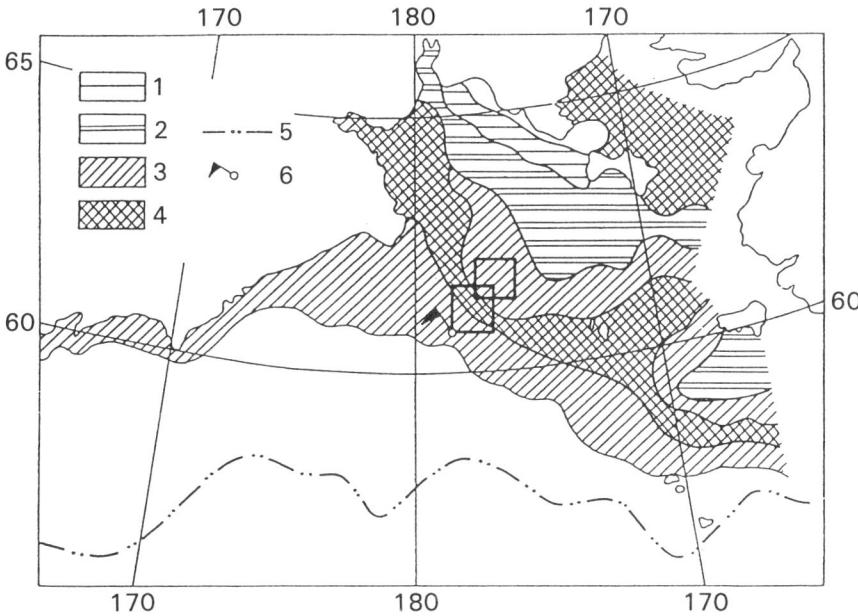

Fig. 7.4. Ice dynamics in the test area of the Bering Sea Experiment, 5–6 March 1973. 1, grey ice; 2, grey-white ice; 3, white ice domination; 4, white ice with moderate first-year inclusions; 5, mean ice propagation boundary in March; 6, weather ship "Priboy" location (Kondratyev et al., 1973a)

To provide the additional ice information, the AARI's aircraft Antonov-24 equipped with the side-looking radar "Toros" was also used in this campaign. With the help of this radar the SLR-images of ice cover and sea roughness were obtained.

Microwave ice cover parameters from the Ilyushin-18 aircraft were obtained as microwave images ($\lambda = 3.2$ cm) and as brightness temperatures (T_B) at the wavelengths 0.8, 1.35, 1.6, 2.0 and 3.2 cm (with varying angle of view of the antenna and polarization). The Convair-900 has the 1.55 cm scanning radiometer, and devices operating at wavelengths 0.8, 0.96, 1.35 and 2.81 cm. An intercomparison of the overlap areas of the USSR and USA microwave images shows good coincidence and the pictures of plynyas and leads in the overlap zones on all days of study. The microwave and radar data of both aircraft gave a good correlation of the distribution of ice concentration over the overlap area.

During the expedition the meteorological and hydrological observations were performed from the weather-ship "Priboy", as well as the ice situation being visually described and the electrophysical parameters of sea ice being measured.

These data were used as the basis to calculate emissivity of different forms of the first-year ice in the Bering Sea in winter (nilas, grey ice, grey-white and white ice)—Fig. 5.6. Calculations demonstrated that the white ice, as the thicker ice, has a maximum of microwave emissivity, and nilas has a minimum value. In the stormy weather when nilas can be covered by saltwater films, the emissivity of this multi-layer structure can be less

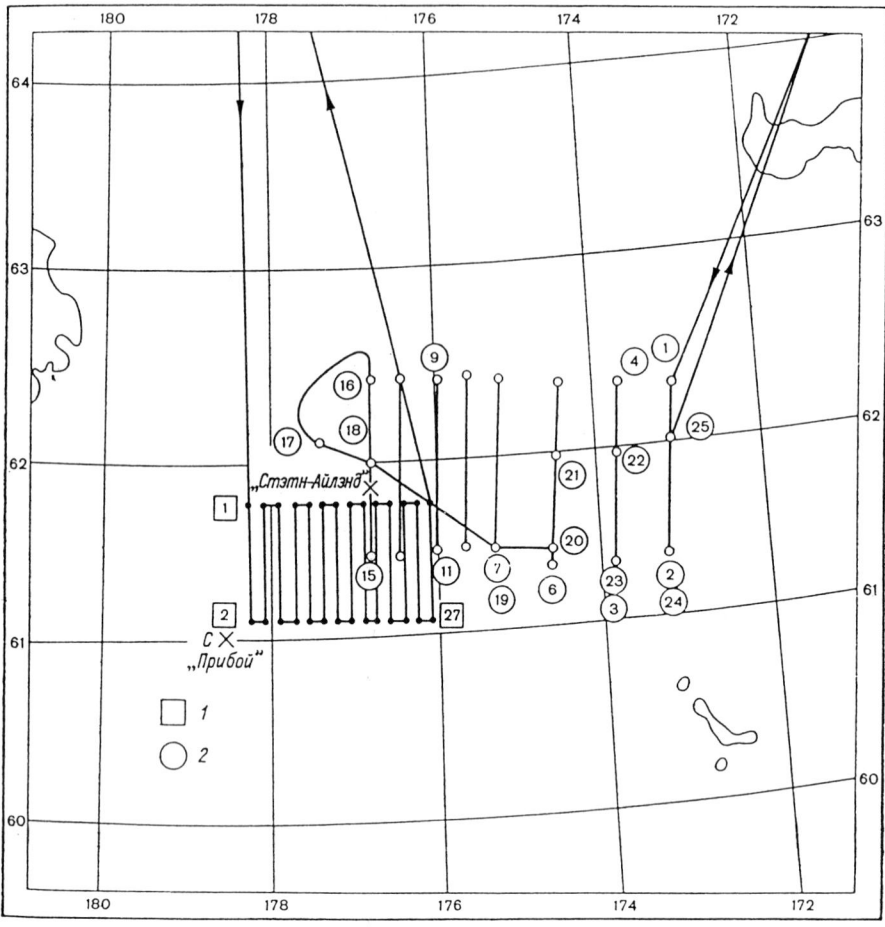

Fig. 7.5. Sub-programme of sea ice study: flight tracks 5 March 1973. 1, Ilyushin-18 passes;
2, Convair-990 passes; (Kondratyev et al., 1973a)

than that of the open water. The results of these theoretical calculations have shown that
the difference in the first-year ice emissivities permits the microwave diagnosis as a
minimum of two ice age gradations for the 0.5–1.5 K s^{-1} radiometer sensitivity level.

An importance result from BESEX is the specified relationship between brightness
temperature and first-year ice solidity ($\lambda = 3.2$ cm)—Fig. 5.7. Considerable scattering of
experimental values in this plot indicates the influence of structure on the indefiniteness
of electrophysical properties and brightness temperatures of the Bering Sea ice cover and
the necessity of an additional study of this phenomenon.

The experience of the "Bering Sea Experiment" has shown also the need for development of the in-flight system of visualization of the microwave information, stronger
sensitivity of radiometers, and providing the digital information suitable for the operational analysis—thematic processing with the use of the in-flight computer.

The principal innovation in the development of the ice mapping complex was the development of an antenna with an electrically scanning beam. To provide the constant resolution on the surface for the coordinate transverse with respect to the direction of the flight, the angle Φ between the planes of flat phased grids was chosen from the relationship (Alexandrov et al., 1986)

$$\sin^2 \frac{\Phi}{2} = \cos^2\left(30° = \frac{\Phi}{2}\right)\sin\left(30° + \frac{\Phi}{3}\right)\cos^2 v_{max} \sin\left(\frac{\Phi}{2} + v_{max}\right) \tag{7.1}$$

where v_{max} is the direction of the edge of the viewing sector.

The scanning system to map the ice cover included the antenna, the high-sensitivity radiometric receiver (0.05 K s^{-1}), the airborne automated scanning serial control, the preliminary processor for the input information, and the video-control apparatus (VCA) to record the false-colour image of emission. The scan lines of preliminarily processed data were recorded on a magnetic disk. Subsequently, the data are further processed to filter out the noise, to calculate the T_B, from the raster with the T_B field, to reduce the T_B to the regular grid, to transform the T_B values to colour gradation numbers, and to record the aircraft coordinates (Kondratyev et al., 1992b).

Figs 7.6 and 7.7 (see the colour section) show the samples of the ice cover microwave images of the Kara Sea obtained from the Ilyushin-18 VCA screen: consolidated pack ice and ice edge, respectively. The right-hand part of the frame contains the false-colour calibration scale. Analysis of the airborne electrically scanning data shows that this instrument permits to map the main arctic sea ice climatic parameter—ice concentration. The state of the ice cover can be defined using indirect indicators of ice age, wind and water surface conditions. The analysis of the microwave observational data shows that changes of ice roughness result in variations of T_B and of the type of its angular dependence. However, in the case of mapping a sharp interface, the ice age can be assessed only qualitatively.

To take this into consideration the characteristics of the antenna system, an external calibration of the radiometric receiver was made for the emission of a homogeneous surface with the emittance properties known—for example, a smooth sea or fresh-water surface with low-varying surface temperatures. Usually for this purpose we used the region of thermally inert part of Lake Ladoga. The calibration temperature is determined during the flight with the use of data of the airborne IR radiometer and thermohygrometer on the vertical distribution of meteorological parameters, as well as the data of aerological sounding.

7.3 AIRBORNE MICROWAVE MAPPING OF THE SEA OF OKHOTSK ICE COVER AND THE TECHNIQUE OF MICROWAVE REMOTE SENSING OF SEA ICE CONCENTRATION

Further development of the MGO Ilyushin-18 scanning microwave system made it possible to outline the water–ice and ice–land interfaces, to map the spatial distribution of sea ice, to estimate in more detail the ice concentration, and to estimate the size of ice floes, polynyas, and leads (Kondratyev et al., 1992b). The airborne microwave data were used to the systematically map the ice climatic parameters because in addition to concentration, it was possible to identify four ice structures in the open water-to-compact ice transition zone: (i) newly formed sea and fresh ice in the ice–water interface zone, (ii) the fast-ice (iii) the ridged and rafted zones of pack and fast-ice (especially in the arctic summer period), and (iv) multi-year and glacier ice fields (especially in the melting period). It was also shown that changes in the ice surface roughness, free water and air inclusions in the ice result not only in variations of the T_B amplitude, they also affect the degree of the T_B angular dependence. The results of these measurements and the corresponding calculations of these phenomena, were combined with similar investigations elsewhere, e.g., Carsey (1992) established the basis for all-weather mapping of the Arctic seas using microwave sensors.

Between 1983 and 1985 three large-scale MGO Ilyshin-18 missions were organized for the microwave survey of the seas and offshore areas of the Arctic Basin. In the winter season, in February–March, an annual series of flights was made along the coordinated routes from the Chuckchi Sea to the Barents and White Seas, with the bases of the aircraft at Anadyr, Cape Schmidt, Magadan, Cherksy, Tiksi, Khatanga, Norilsk, Amderma, and Murmansk. The microwave aircraft survey was supported by the spaceborne instruments in the microwave (active and passive), visible, and IR regions. All the stages of the three-year expedition were coordinated with the ground-truth studies of ice parameters.

In the far east, at the Bering, Okhotsk and Japan Seas and in the Pacific Ocean (the Kuril–Hokkaido region), the state of the ice and sea surface was studied in the coordinated periods using the research ships "Academician Shirshov", "Priliv", and "Priboy". In the northern part of Cape Schmidt the aircraft experiments on the microwave sounding of sea ice were accompanied by the radar survey of ice with the use of the ground-based radar station. To monitor the ice dynamic the Ilyushin-18 crew included also an AARI's expert on the visual ice analysis which made it possible to check the ice phenomena more detailing. The airborne observations were complemented by multi-spectral satellite data. Each stage of the three-year project was coordinated with field studies, providing surface validation.

Scanning radiometer measurements were used to estimate the ice concentration—see Fig. 7.8 which shows the histograms of T_B distribution at $\lambda = 2.05$ cm (14.6 GHz) for two transections extending for 35 and 80 km, in the southeast margin of the Sea of Okhotsk near the Kuril Islands. The first histogram was obtained from the data of the computer processing, 19 2000 values of T_B, and the second histogram from 44 800 values. As can be seen from the data, both areas are characterized by the bimodal distribution of T_B. The first mode corresponds to the open water and initial stages of ice formation, the second mode corresponds to the developed first-year ice. The values of T_B obtained for different kinds of first-year sea ice, agree with the data of numerical modelling.

The analysis of the histograms shows also that for the region of the Strait of Freeze (46° N, 149° E—Fig. 7.8(a)) the share of sea surface, both open and covered with initial types of ice, is higher than for the region of the Strait of Katherine (45° N, 147° E—Fig. 7.8(b)). The revealed specific ice distribution coincides with the data of the 10-day map of ice distribution in the Sea of Okhotsk compiled from the results of visual survey.

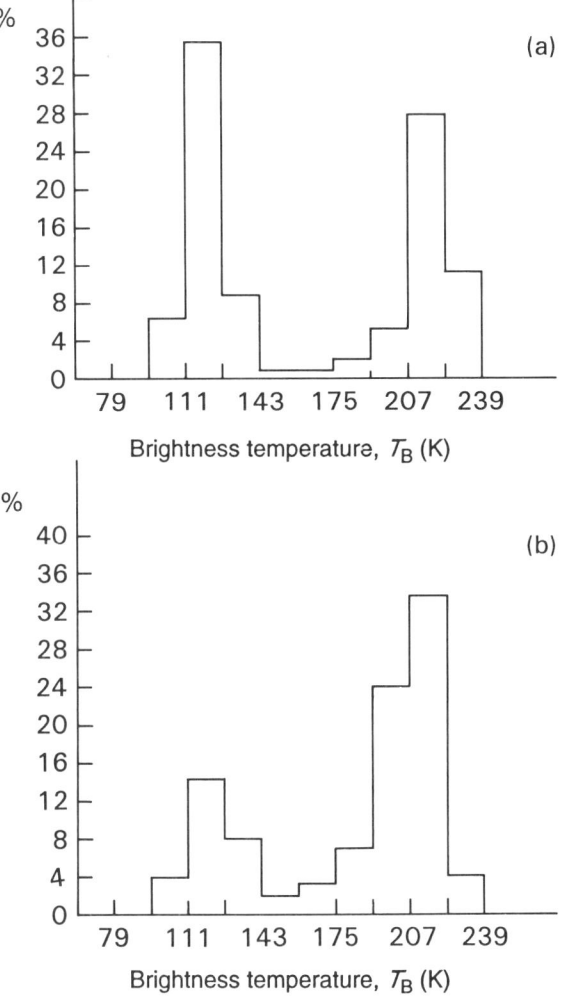

Fig. 7.8. Histograms of the brightness temperature of two ice side of different concentration, $\lambda = 2.05$ cm, Okhotsk Sea: (a) Strait of Freeze, (b) Strait of Katherine (Melentyev and Aleksandrov, 1991)

The next step in parameter retrieval is the quantitative estimation of ice concentration. The 14.6 GHz measurements over the Sea of Okhotsk revealed that water surface T_B ranged from 110–125 K. These measurements agree well with T_Bs calculated using realistic values of sea roughness and atmospheric state. Measurements were then made over extended, homogeneous fields of first-year ice (from visual assessment); the mean $T_B = 221$ K. The T_Bs for open water and ice were then used to estimate ice concentration from the histograms for the entire transection. Based on the emissivity of first-year ice, a threshold level (T_{Bcrit}) was selected. The ice concentration is retrieved by dividing the number of values $> T_{Bcrit}$ by the number of values $< T_{Bcrit}$.

The uncertainties in T_B estimation due to errors in reducing the scanning radiometer data to nadir, result in a scatter in the ice concentration estimation not more than 5%. Because the antenna field-of-view can incorporate both water and sea ice of different types, the probabilities of receiving emission from different surfaces largely determine the character of the T_B distributions. This can limit the effectiveness of ice concentration estimation and ice classification based on T_B frequency distributions.

The estimate of the Sea of Okhotsk ice concentration along the Strait of Freeze from the data of microwave survey is 48%; along the Strait of Katherine it is 71%, which agrees well with the data of the ten-day map compiled from the ice survey by the I. Sakhalin Hydrometeorological Service.

7.4 SEA ICE THICKNESS ESTIMATION FROM THE LOW-FREQUENCY MICROWAVE SOUNDING: RESULTS OF THE AIRBORNE SURVEY IN THE SEA OF OKHOTSK AND CHUCKCHI SEA

The important problem of microwave remote sensing is an evaluation of ice age and ice thickness as other important ice climatic parameters. The centimetre and millimetre parts of the microwave region are known to have the limits in determination of these meteoelements. These sea ice parameters could be retrieved using low-frequency microwave radiometry, but the satellite measurements at decimetre wavelengths are limited at present due to the unwieldiness of the large antenna required. For this reason the experience of Ilyushin-18 microwave survey at $\lambda = 18.1$ and 35.1 cm (1.7 and 0.9 GHz, respectively) in the arctic and sub-arctic regions is interesting.

The data from Ilyushin-18 microwave measurements have shown some prospects for the determination of first-year sea ice thickness, related to the depth of the sensed layer, which increases at this wavelength.

Fig. 7.9 shows the 8 February 1983 distribution of the Sea of Okhotsk ice cover near Magadan (60° N, 151° E). The ice concentration estimated using the histogram computer processing technique described above is 100%. The ratio of the older (thick) ice to the younger (thin) ice with, respectively, higher and lower T_B values, has been marked in the denominator of a "circle" index (its left- and right-hand parts).

The following ice types (with associated thickness) were identified: dark nilas (5 cm, $T_B = 110$–118 K), fields of grey ice (10–15 cm), fields of grey-white ice (15–30 cm), and thin first-year ice (30–70 cm). In the spring of 1985, a survey of the same region was repeated, revealing the presence of two other types of ice in the northeastern part of the region: moderate first-year ice (70–120 cm) and thick first-year ice (> 120 cm, $T_B = 215$–220 K).

Analysis of these microwave measurements and visual observations shows that 1.7 GHz brightness temperatures enable one to distinguish between thick first-year ice (120–150 cm), prevailing in this region at this time, and moderate white ice (≈ 70 cm). The moderate white ice, connecting the thick first-year ice floes, formed in the areas where polynyas had earlier been; the moderate white ice fields are of the order of several hundred square kilometres. The mean 1.7 GHz T_Bs were 242 K for thick first-year ice, and 234 K for moderate white ice.

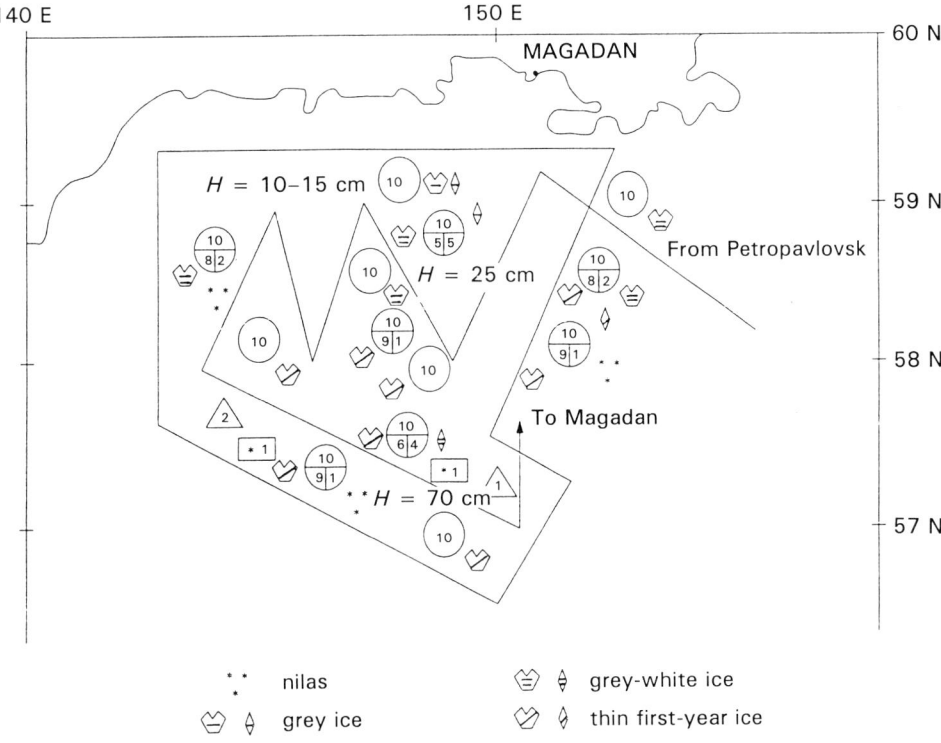

Fig. 7.9. Map of sea ice parameters derived from 1.66 GHz microwave survey of the Sea of
Okhotsk ice cover (Kondratyev and Melentyev, 1994)

In a separate investigation, multi-frequency measurements were made in the Chuckchi
Sea on 28 February 1984. The three-channel (1.7, 2.2, and 2.7 GHz) measurements in the
decimetre range made it possible to identify five levels of first-year ice thickness. A
comparison of the data of the airborne microwave and visual observations has shown that
measurements at $\lambda = 18.1$ cm enable one to distinguish between thick first-year sea ice
(120–150 cm) prevailing in the region at this time, and moderate white ice 60–90 cm
thick. The moderate white ice, connecting the floes of the thick first-year and multi-year
ice, has been formed in places of drifting when the polynyas became frozen. The area of
the moderate white ice inclusions from the data of the airborne survey was of the order of
several hundred square metres (usually, the moderate white ice was free of snow, blown
away by winds from its smooth surface). The mean T_B values at $\lambda = 18.1$ cm constituted
242 K for thick first-year ice, and 234 K for moderate white ice. The T_Bs correspond to
the ice classification summarized in Table 7.1.

The considerable variety of the age types of ice, revealed by the airborne microwave
survey, confirms the fact that the Chuckchi Sea is a specific arctic basin in which the
trend of variability of the thermal state of surface waters drastically differs from that of
the adjacent seas. This is known to be connected with the special wind regime in this
region of the Arctic (Zakharov, 1981).

Table 7.1. Brightness temperatures (1.7 GHz) of the ice in the Chuckchi Sea,
28 February 1984

Type of sea ice	Multi-year > 150 cm	One-year (thick)	White ice (moderate)	Grey-white	Nilas	Water 0°, 30‰
$\lambda = 18.1$ cm	245	235–245	225–235	210–235	105–120	93–96

In these investigations and others, e.g., Comiso (1994), ice thickness is inferred through ice classification and associated ice thickness. There is at present no microwave system that can directly retrieve this parameter (Wadhams and Comiso, 1992). Since ice thickness is important climatologically and for understanding the ice dynamics, a reliable retrieval of ice thickness from microwave data would be a major breakthrough. The relative lack of similar research using low microwave frequencies is largely due to the practical restrictions of the antenna size required for low-energy measurements. That is, even if low-frequency microwave measurements exhibit a strong dependence on the ice thickness, the implementation of such a sensor system appears unfeasible.

7.5 ACTIVE AND PASSIVE MICROWAVE SENSING OF FAST-ICE IN THE EAST SIBERIAN SEA

The experience of decoding the combined radar and microwave data has demonstrated that the use of both techniques gives versatile information. The usefulness of combining both active and passive microwave measurements was investigated in the East Siberian Sea fast-ice zone in February 1983.

The low-altitude (400 m) Ilyushin-18 mission started over land (15–20 km from the coast) and ended at a distance of 60 km over the sea ice in long Strait, connecting the East Siberian and Chuckchi seas. The subsequent flight transections, including those in the opposite direction, were made 10 to 20 km eastward. The passive microwave measurements were accompanied by active microwave observations from the ground-based meteorological radar station (MRS-2) at Cape Schmidt (68° N, 179° W). The radar station is on an islet 60 m from the coastline. The height of the 9.4 GHz, circular-scanning antenna system was 7 m above sea level, with an antenna angle of 0.7°, permitting a range of 16 km under normal conditions of refraction.

The presence of ice can be detected with radar by differences in surface roughness, which influences the radar return. The MRS-2 radar identifies best the older, deformed ice, which has a high return. This is explained by the fact that at the initial stage of ice formation the ice cover is a smooth surface covered with a thin snow layer. The indirect indications were used to identify the ice of initial forms. So, for example, in the case of rough seas, the signal is observed as the homogeneous background emission. With the initial ice present, the indicator shows dark bands stretched along the coastline. The young ice, being rather plastic, damps the sea roughness, the radio signal is not reflected from the rough sea surface, and the VCA screen reveals the dark bands. But in the spring and summer periods, round fragments of near-shore ice with the smooth snow cover can give similar effects. Against the radio-echo of the ice with a 70–100% concentration, this ice shows itself as dark spots which can be identified from some indirect indicators.

The 4 February 1983 MRS-2 data revealed the presence of near-shire fast-ice, with a series of ice ridges oriented mainly along the coastline (Fig. 7.10). The ridges were formed in clearings beyond the near-shore ice, with subsequent compression in the two-month period from December 1982 to the time of survey. The first zone, resulting from the compression of the near-shore ice, about 2 km wide, is 2–2.5 km from the coast. The second zone is at a distance of 8 km. The third zone detected by the radar is 1 to 5 km wide, 13 km from the coast; it consists of 4–5 ridges formed at different times.

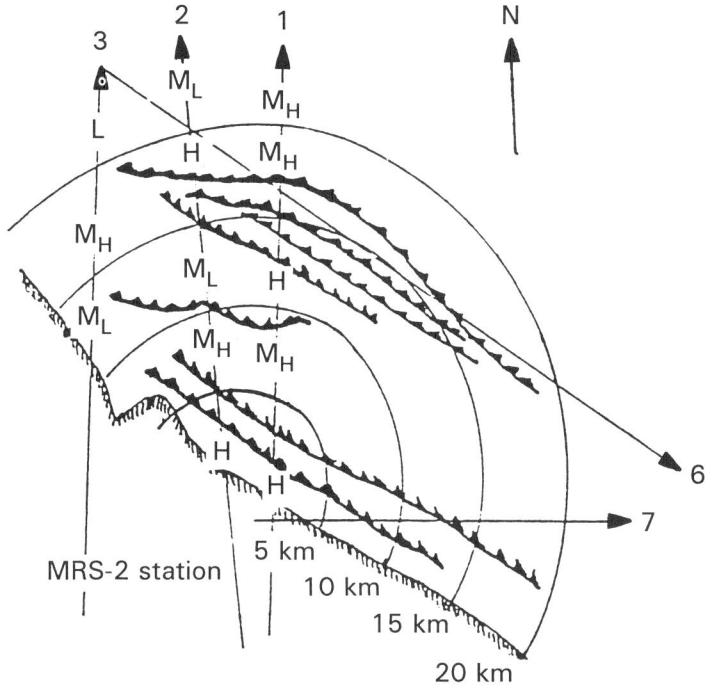

Fig. 7.10. Schematic of combined passive microwave and radar sensing in the Cape Schmidt region. Flight transections are indication with lines. Ice ridges are indicated with serrated lines. Codes refer to microwave-estimated ice concentrations: L < 25%; M_L = 20–40%; M_H = 40–60%; H > 60%; (Kondratyev and Melentyev, 1994)

An attempt was made to find out the relationship between the passive microwave T_B variability (standard deviation of brightness temperature, (σT_B) of homogeneous, extended ice areas and variations in the first-year ice with ridges. Four levels of ice concentration are considered: (i) < 20% (L—low); (ii) 20–40% (M_L—moderate/low); (iii) 40–60% (M_H—moderate/high); and (iv) 60–90% (H—high), corresponding to the following values of σT_B at 1.7 GHz: ±1 K; ±2 K; ±3 K; and ±4 K. The ice concentration in the region of the Ilyushin-18 mission on 4 February 1983 varied from 20% to 90%, in agreement with the MRS-2 data.

The use combined active and passive microwave data has shown that these techniques supplement each other, and can provide information about the sea ice conditions needed to study Arctic climate and to navigate in the Arctic coastal regions. Indeed, combined satellite passive and active microwave systems are beginning to be used for such purposes in the Siberian marginal seas (Johannessen et al., 1992).

7.6 MICROWAVE REMOTE SENSING OF FRESH-WATER ICE AND SNOW CLIMATIC PARAMETERS

In winter, vast expanses of northern Siberia, largely comprising tundra, are generally snow-covered, with frozen rivers and lakes. These high-latitude features were also investigated in winter using high-altitude aircraft survey. The microwave properties of such features were investigated using the MGO Ilyushin-18 aircraft (Kondratyev et al., 1992b). The 4 February 1983 flight was started at Anadyr (64° N, 178° E). Passive microwave measurements ($\lambda = 18.1$ cm) were made over the region of high-mountain tundra north of Anadyr to Cape Schmidt. The terrestrial part of the transection of the survey revealed the ability to identify the areas of fresh-water ice (vast, frozen lakes in the tundra), with mean $T_B = 206$ K, as well as lake ice fractures. The continental snow-covered tundra is characterized by mean $T_B = 230$ K, with standard deviation $(\sigma T_B) = \pm 6$ K, which is somewhat higher than the respective σT_B for the sea ice cover in the adjacent Chuckchi Sea (Long Strait region).

Remote sensing of ice during the melt period is also of practical importance. For example studies of the thermodynamic and radiative regimes of water basins, atmosphere–ice–water interaction processes, the ice regime of the inland water basins, and assessments of river runoff and flooding. A technique has been proposed (Kondratyev et al., 1985, 1992b) to retrieve the characteristics of melting ice using microwave radiometers. This approach uses a cross-track scanning antenna system, thus making use of the angular dependence of emittance. The angular emittance characteristics of ice and water are used to assess the state of the melting ice. This technique was developed using a series of 14.6 GHz passive microwave measurements made in April 1983, at altitude 2900 m above the high-mountain Lake Sevan in the Caucasus. Flights were made along the longitudinal axis of the lake and around its margins.

The results for one transection are shown in Fig. 7.11. As seen from these data, the distribution of the lake ice cover emission is heterogeneous. The highest T_Bs correspond to the ice cover in the extreme southeastern part of the lake; here the ice appears white, generally smooth with some ridging. Several dark lineaments and streaks 3–5 m wide are seen, and are also easily identified as inhomogeneities on the on-board real-time T_B plot; these fractures and streaks are marked with vertical dashes of different length, characterizing different widths of streaks (Fig. 7.11). A polynya observed beyond the fast-ice was also detected on the T_B plot.

Smaller ice fields of similar structure can be identified from the visual and microwave survey over the Great Sevan in the part adjacent to the strait separating it from the Small Sevan. This springtime fresh-water ice, conditionally identified as white ice, has a mean

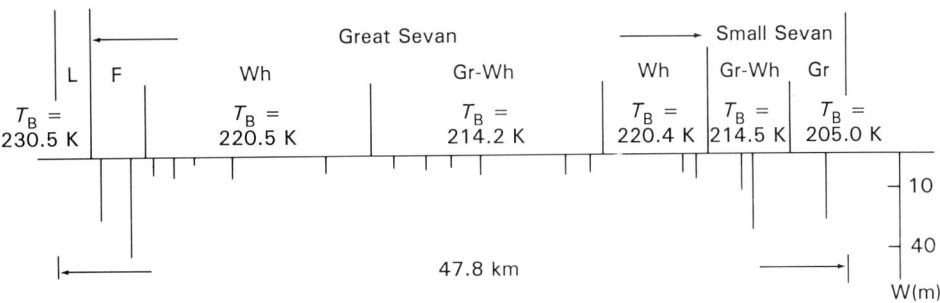

Fig. 7.11. Distribution of Lake Sevan ice types and brightness temperatures measured with an airborne radiometer. Surface codes are: land (L), fast-ice (F), white ice (Wh), grey-white ice (Gr-Wh), and grey ice (Gr) (Kondratyev and Melentyev, 1994)

$T_B \approx 220$ K. Most of the Great Sevan was covered with darker ice, classified as grey-white ice. As the aircraft passed over the Small Sevan, a large band of disturbed ice was seen; it was classified as grey-white and grey ice. The mean T_Bs for these ice types are 214 K and 205 K, respectively. The flight over this region also revealed several fractures with a corresponding T_B decrease.

Horizontally polarized measurements at 14.6 GHz were made on the following three levels of melting, spring ice: (i) small; (ii) moderate; and (iii) large (the ice has nearly disintegrated). The following values of the parameter $\Delta T_B(\Theta_{0-50})$ were ascribed to these levels: (i) 15 K; (ii) 15–25 K; (iii) 25–35 K. Values of $\Delta T_B(\Theta_{0-50})$ >35 K correspond to the open-water areas. Fig. 7.12 is an example of the Great Sevan T_Bs in the 0–50° scanning regime. The minimum $\Delta T_B(\Theta_{0-50})$ refers to the fast-ice; white ice and grey-white ice differ in both the absolute value of emission ($T_{Bw} > T_{Bg/w}$) and $\Delta T_B(\Theta_{0-50})$. Grey-white ice has a greater liquid water content, and a correspondingly higher value of $\Delta T_B(\Theta_{0-50})$.

The results of these airborne investigations of land, snow, and fresh-water ice have demonstrated that various surface characteristics can be assessed using their microwave emission properties. These characteristics include the ice structure and the presence of liquid water in the ice, which makes it possible to assess indirectly such parameters as its density. It is also possible to identify fractures, streaks, polynyas, fresh-water, ice–land interfaces, snow-covered land areas, and snow-free land areas. Here, the angular dependence of microwave emission is useful, as suggested by laboratory measurements.

7.7 AN EXPERIENCE OF SIDE-LOOKING RADAR (SLR) AND PASSIVE MICROWAVE SOUNDING OF LAKE AND SEA ICE

In studies of ice parameters multi-level experiments were carried out with the use of aerospace means of remote active and microwave sounding "Kosmos-1500" and airborne SLR complexes of various radar instruments (Gavrilenko et al., 1986; Kalmykov et al., 1984) with simultaneous surface *in situ* measurements of hydro- and radiophysical characteristics of ice.

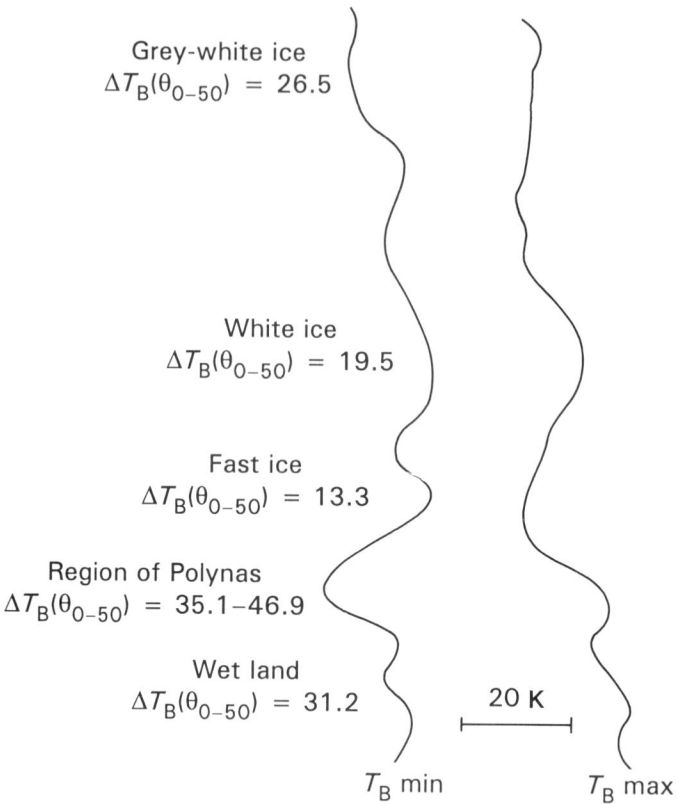

Fig. 7.12. Along-transection plot of Lake Sevan ice types and associated microwave parameters. $\Delta T_B(\Theta_{0-50})$ is the mean difference in horizontally polarized brightness temperature as measured between 0°–50° (Kondratyev and Melentyev, 1994)

Airborne remote sensing was preceded with studies of the mechanism of the formation of radar signal scattered by freshwater ice. For this purpose an X-range scatterometer mounted on a special platform at a height of about 4 m above the ice surface was used. Backscattering from ice was measured in the range of incidence angles 0–70°, the distance to the ice surface under study being kept constant. Surface experiments have shown that for fresh-water ice characterized by low absorption ($\leqslant$ 1 dB m–1) and a small number of volume scatters ($\leqslant$ 1%), a radar signal is mainly formed by reflections from the upper and lower boundaries of ice.

Complex multi-level observations on Lake Onega (Kondratyev et al., 1988b) were carried out using the following technique. The satellite radar information was used to assess the state of ice cover over the whole lake area. At the quasi-synchronous time the ice area was mapped with the airborne SLR complex as well as visual ice reconnaissance was made. *In situ* measurements of the parameters of ice and snow cover were taken by the helicopter which landed at various points. At these helicopter landings observations were made of the type of ice, its thickness, surface roughness, the height and density of snow cover, the temperature was measured of air, snow surface, and snow–ice interface. Ice cores were taken to determine the density and porosity of ice.

The survey was made during the period of solid ice cover without open water; the snow was dry and fine-grained. One can see two areas of lake ice cover of different reflexivity in the western and eastern parts of the lake where the ice is rough. The SLR-range enables the recognition of the history of the formation of the ice cover. Westerlies prevailed in the period of lake freezing.

The SLR space-derived radar images reveal two areas of the lake ice cover with different shapes of the upper surface of the ice due to spatial resolution insufficient for characteristic inhomogeneities of the lake ice cover.

A detailed study of the ice cover can be made using SLR-images drawn from the aircraft measurements. A fragment of media map is shown in Fig. 7.13.

In the eastern part of the lake numerous individual ice fields driven by westerlies during the period of ice formation froze together to make a solid ice cover. The boundaries of ice fields are contoured by the zones of pack ice (light strips). In the western part the ice surface is relatively smooth. The angular dependences of the backscattering coefficient σ° for slightly rough (1) and hummocky (2) sites were determined from the data of the airborne SLR.

In synchronous studies of the characteristics of the state of freshwater ice individual parts of ice and water are reliably determined from the Ku-range radiometric information, whereas radar images identify individual parts of smooth ice as open water. Microwave and infrared airborne data can be used to assess the snow-cover parameters.

In contrast to fresh-water ice, in the case of sea ice three mechanisms participate in the formation of radar signal—scattering by rough boundaries, absorption, and volume scattering within the ice layer. Depending on the age of ice, on the ice-formation conditions, and on meteorological parameters, the contribution of each of the mechanisms can vary, which leads eventually to considerable variations in the σ° parameter of ice cover. As observations with the airborne radiophysical complex have shown, 5 to 6 age gradations of sea ice can be identified in winter as well as cracks and leads in the ice cover (Gavrilenko et al., 1986).

Experience of using the "Kosmos-1870" SAR suggests the conclusion that this system is also a valuable source of operational information about the state of ice cover (Kondratyev et al., 1987). Radar images permit the identification of important ice climatic parameters: ice–water interface, leads and parts of thin ice among first-year and multi-year ice, as well as large fields of multi-year ice against first-year ice (Kalmykov et al., 1984). The data of ice survey obtained by the "Kosmos-1870" SAR enabled one to recommend optimal ship routes with the use of natural leads in the ice cover.

Scattering of microwaves by fresh-water ice is caused by reflections from the upper and lower surfaces of ice, scattered on the rough upper boundary.

As demonstrated by experience the combined aircraft X-range SAR data and scanning radiometer Ku-range data make it possible to identify 5 to 6 age gradations of sea ice, and to find and identify leads, structure and parameters of the roughness of freshwater and sea ice cover.

Experience of microwave passive sounding of lake and sea ice, both in winter and during the period of ice destruction, has shown the possibility of retrieving its various climatic characteristics: structure, the presence of open water, enabling one to assess the density of ice, its solidity and destruction; it has also shown the possibility of determining the open water parts, and the ice–snow-covered land interface.

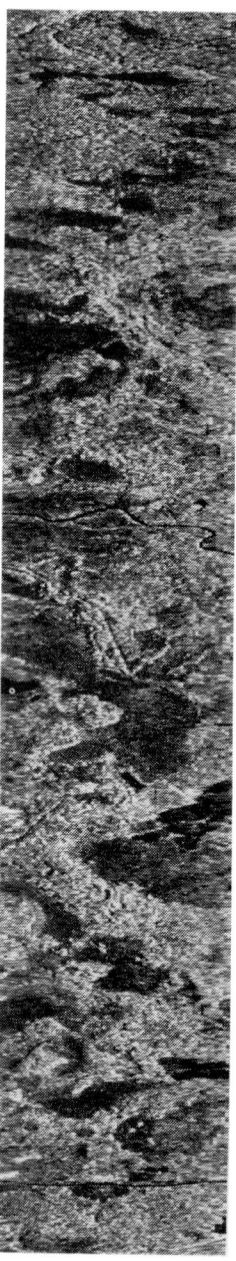

Fig. 7.13. Fragment of the SLR-image of fresh ice (Onega Lake, 18 February 1988; $\lambda = 3.15$ cm)
(Kondratyev et al., 1987)

8

Conclusion

It is quite clear from what has been discussed in the book that high latitude environmental dynamics deserves special attention. On the other hand, it is equally obvious that there are many unsolved problems whose solution is badly needed because of the rapidly progressing economic development at high latitudes (particularly in the Arctic, of course). In Chapter 1 a number of on-going and forthcoming programmes of high latitude environmental studies have been discussed. To avoid repetition, we shall formulate very briefly the most important conclusions:

(1) There are more than enough programmes of future research in the high latitude regions, but what is urgently required is efficient coordination of the existing programmes and, in the ideal case, an integrated programme of polar environmental research. Probably, the Arctic System Science: A Plan for Integration (ARCSS) will be able to serve as a prototype of an integrated programme. The task of integration is especially important because it has been clearly recognized at the present time that environmental and biosphere processes are closely coupled.

(2) There is a huge, but still insufficient observation data bank for the high latitude environment. A much better substantiation and accomplishment of an optimized environmental observing system is necessary, however, which should be based on clearly defined requirements for various kinds of observations, with special emphasis on satellite monitoring and focused field experiments.

(3) Since polar regions are precisely those parts of the globe where the solar activity impact may be most pronounced, this aspect of environmental dynamics should receive more attention.

References and bibliography

Aagard K., Carmack E.C. (1994). The Arctic ocean and climate: a perspective. Geophysical Monograph 85. AGU, Washington D.C. p.5–20.

Abelson P.H. (1989). The Arctic: a key to world climate. Science, Vol.243, N4893, p.104–113.

Ackerman T.R., Valero F.P.J. (1984). The vertical structure of arctic haze as determined from airborne net-flux radiometer measurements. Geophys. Res. Lett., Vol.11, N5, p.469–472.

Ackerman S.A., Eloranta E.W., Grund C.J., Knutsen R.O., Revercomb H.E., Smith W.L., Wylie D.P. (1993). University of Wisconsin. Cirrus Remote Sensing Pilot Experiment. Bull. Amer. Meteorol. Soc., Vol.74, N6, p.1041–1051.

Adamenko V.N., Kondratyev K.Ya., Sinyakov S.A. (1991a). Deposition of metals from the atmosphere at the North Pole compared to background results of the northwestern USSR. Proc. Int. Conf. on the Role of the Polar Region in Global Change (Fairbanks, June 11–15, 1990). Geophys. Conf. Univ. of Alaska, Fairbanks. Vol.II, p.716–719.

Adamenko V.N., Kondratyev K.Ya., Sinyakov S.A. (1991b). Deposition from the atmosphere of metals at the North Pole as compared to background regions of north-western Europe. Izv. All-Union Geogr. Soc., Vol.123, issue 4, p.316–323 (in Russian).

Adriani A., Deshler T., Gobby G.P., Johnson B.J., Di Donfrancesco G. (1992). Polar stratospheric clouds over McMurdo, Antarctica, during the 1991 spring: lidar and particle counter measurements. Geophys. Res. Lett., Vol.19, N.17, p.1755–1758.

Aerosol and Climate. Ed. by K.Ya. Kondratyev (1991). Leningrad, Gidrometeoizdat. 542 pp.

Alekseev G.V. (1994). The influence of polar oceans on interannual climate variations. Geophysical Monograph 85. AGU, Washington, D.C. p.327–336.

Alekseev G.V., Podgorny I.A., Svyaschennikov P.N. (1991). Oscillations of ocean warming impact on global climate. Doklady USSR Acad. Soc., Vol.320, N1, p.70–73 (in Russian).

Alexandrov V.Yu., Brovirova T.Yu., Melentyev V.V. (1986) Technique and some results of processing the ice-water emission contrasts from the airborne scanning radiometer. Trudy GGO. Issue 509, p.162–178 (in Russian).

Angell J.K. (1993). Reexamination of the relation between depth of the Antarctic ozone hole, and equatorial QBO and SST, 1962–1992. Geophys. Res. Lett., Vol.20, N15, p.1559–1562.

Arcorne S., Delanay A.J., Sellman P.V. (1979). Detection of Arctic water supplies with geophysical techniques. U.S. Army Cold Region Laboratory, Hanover, NH. CRReL Report 7915.

Arctic Air Pollution (1986). Ed. B. Stonehouse. Cambridge University Press, Cambridge. 328 pp.

Arctic Climate System Study (ACSYS). Initial Implementation Plan. (1994). WCRP-85 (WMO/TD, N627), Geneva, Switzerland. 66 pp.

Arctic Research for an Arctic Nation (1989). U.S. Arctic Research Commission, Washington, D.C. 39 pp.

Arctic System Science. Ocean–Atmosphere–Ice Interactions. Joint Oceanogr. Inst. Inc. Washington, D.C. (1990). 132 pp.

Assessment of the ice cover characteristics with microwave remote sensing techniques (1978). Leningrad, MGO. Report. no. 78032305.

Baker D.J. (1981). Ocean climate effects in polar regions. Techn. Studies Related to the Development of a System for Ocean Climate Monitoring, Vol.2, Contribution from Consultants. NCAR, Boulder, Co. p.7–8.

Barber D.G., Manore M.J., Agnew T.A., Welch H., Sonlis E.D., De Drew E.F. (1992). Science issues relating to marine aspects of the cryosphere: implications for remote sensing. Canad. J. Remote Sens. Vol.18, N1, p.46–55.

Barnett T.P., Dumenil L., Schlese C., Roeckner E., Latif M. (1989). The effect of Eurasian snow cover on regional and global climate variations. J. Atmos. Soc., Vol.46, p.661–685.

Barrie L.A. (1986) Arctic air pollution: an overview of current knowledge. Atmos. Environm. Vol.20, N4, p.643–663.

Barrie L.A., Den Hartog G., Bottenheim J.W., Landsberger S. (1989a) Anthropogenic aerosols and gases in the lower troposphere at Alert, Canada, in April 1986. J. Atmos. Chem., Vol.9, N1–3, p.101–128.

Barrie L.A., Olson M.P., Oikawa K.K. (1989b). The flux of anthropogenic sulphur into the Arctic from mid-latitudes in 1979–90. Atmos. Environm., Vol.23, N11, p.2505–2512.

Barry R.G. (1990). Evidence of recent changes in global snow and ice cover. Geo. Journal, Vol.20, p.121–127.

Barry R.G. (1991). Observational evidence of changes in global snow and ice cover. In M.E. Schlesinger (ed.), Greenhouse-Gas-Induced Climatic Change: A Critical Appraisal of Simulations and Observations. Elsevier, Amsterdam, p.329–345.

Barry R.G., Crane R.G., Schweiger A., Newell J. (1987). Arctic cloudiness in spring from satellite imagery. Journal of Climatology, Vol.7, p.423–451.

Barry R.G., Maslanik, J., Steffen K., Weaver R.L., Triosi V., Cavalieri D.J., Martin S. (1993a). Advances in sea-ice research based on remotely sensed passive microwave data. Oceanography, Vol.6, N1, p.4–12.

Barry R.G., Serreze M.C., Maslanik J.A., Preller R.H. (1993b). The Arctic sea ice–climate system: observations and modelling. Rev. Geophys., Vol.31, p.397–422.

Barry R.G., Fallot J.M., Armstrong R.L. (1994). Assessing decadal changes in the cryosphere: Eurasian snow cover. Proc. Fifth Symposium on Global Change Studies. American Meteorological Society, Boston, p.148–155.

Basharinov A.E., Gurvich A.B. (1970) Studies of the Earth microwave emission and that of the atmosphere using the Kosmos-243 satellite. Vestnik AN SSR, N10, p.37–41 (in Russian).

Basharinov A.E., Gurvich A.S., Egorov S.T. (1974) Radio-emission of the Earth as a planet. Moscow, Nauka Press. 188 pp. (in Russian).

Baumgartner M., Apfli G. (1994). Towards an integrated geographic analysis system with remote sensing, GIS and consecutive modelling for snow cover monitoring. Int. J. Remote Sensing, Vol.15, N7, p.1507–1517.

Bekoryukov V.I., Borisov Yu.A., Zvyagintsev A.M., Kruchenitsky G.M., Perov S.P., Rudakov V.V. (1994). Observations of two ozone "mini-hole" over Europe in the early winter of 1992–1993. Izv. RAS. Physics of the Atmosphere and Ocean, Vol.30, N6, p.807–811 (in Russian).

Bespalova E.A., Rabinovich Yu.I., Sharkov E.A., Shiriajeva T.A., Etkin V.S. (1976) Study of the processes of ice formation from the aircraft microwave emission measurements. Meteorology and Hydrology, N12, p.68–72 (in Russian).

Bizzarri B., Tomassini C. (1976). Retrieval of information from high resolution images. COSPAR, Proc. Symp. on Meteorol. Observ. from Space: their Contribution to the FGGE. Philadelphia. p.140–144.

Bjork G. (1992). On the response of the equilibrium thickness distribution of sea ice to ice export, mechanical deformation, and thermal forcing with application to the Arctic Ocean. J. Geophys. Res., Vol.97, NC7, p.11287–11289.

Bjork E. and Johannessen, O.M. (1994). Sea ice concentration derived from SMMR and SSMI: parameter retrieval and algorithm comparison. Proceedings European Symposium on Remote Sensing, 26–30 September, 1994, Rome.

Blanchet J.-P. (1989). Toward estimation of climatic effects due to Arctic aerosols. Atmos. Environm., Vol.23, N11, p.2609–2626.

Blanchet J.-P., List R. (1983) Estimation of optical properties of arctic haze using a numerical model. Atmos. Ocean., Vol.21, N4, p.444–465.

Boden T.A., Kaiser D.P., Sepansky R.J., Stoss F.W. (1994). Trends '93: A Compendium of Data on Global Change. CIDIAC, Oak Ridge, TN. 984pp.

Bodhaine B.A. (1989). Barrow surface aerosol: 1976–1986. Atmos. Environm., Vol.23, N11, p.2357–2370.

Bodhaine B.A., Dutton E.G. (1993). A long-term decrease in Arctic haze at Barrow, Alaska. Geophys. Res. Lett., Vol.20, N10, p.947–950.

Bogorodsky V.V., Khokhlov G.P. (1975) Electric properties of the Bering Sea near-shore ice at the frequency 10 GHz. Soviet–American "Bering Sea" Experiment. Proc. Final Symp. on Results of the Joint Sov.–Amer. Expedition. Leningrad, Gidrometeoizdat. p.234–270 (in Russian).

Bogorodsky V.V., Khokhlov G.P. (1978). Relationship between the microwave electric parameters of the upper layers of the arctic drifting ice, its temperature and salinity. Trudy AANII, Vo.359, p.13–19 (in Russian).

Bogorodsky V.V., Martynova E.A. (1978). The own thermal emission of the snow-ice cover of the arctic seas. Leningrad, Gidrometeoizdat. 39 pp. (in Russian).

Bogorodsky V.V., Oganesian A.G. (1987). The penetrating radar sounding of sea ice with the digitally processed signals. Leningrad, Gidrometeoizdat. 343 pp. (in Russian).

Bogorodsky V.V., Gusev A.V., Khokhlov G.P. (1971). Freshwater ice physics. Leningrad, Gidrometeoizdat. 227 pp. (in Russian).

Bogorodsky V.V., Kondratyev K.Ya., Rabinovich Yu.I., Shulgina E.H. (1976). Microwave remote sensing of sea surface with oil slicks. Trudy GGO, Issue 371, p.22–36 (in Russian).

Bogorodsky V.V., Kozlov A.I., Tuchkov L.T. (1977). Microwave emission of land covers. Leningrad, Gidrometeoizdat. 224 pp. (in Russian).

Bogorodsky V.V., Kanareikin D.B., Kozlov A.I. (1981). Polarization of microwave scattered and own emission of land covers. Leningrad, Gidrometeoizdat. 279 pp. (in Russian).

Bogorodsky V.V., Darovskikh A.H., Spitsyn V.A., Khokhlov G.P. (1983). Study of microwave emission of one-year arctic ice covers with account of real distributions of its thickness, temperature, and electric parameters. Trudy AANII, Vol.379, p.26–35 (in Russian).

Bojarsky V.I. (1983). Reflection and propagation of radar signals during the sea ice sounding. Leningrad, IAA. Master's thesis (in Russian).

Bojkov R. (1993). Record low total ozone during Northern winters of 1992 and 1993. Geophys. Res. Lett., Vol.20, N13, p. 1351–1354.

Bordonsky G.S. (1990). Thermal emission of the freshwater basins ice cover. Novosibirsk, Nauka Publ. 104 pp. (in Russian).

Bordonsky G.S., Krylov S.L., Polyakov S.V. (1992). Specific features of radiobrightness of fresh ice cover with gas inclusions. Studying the Earth from Space, N5, p.13–21 (in Russian).

Bourke R.H., McLaren A.S. (1992). Contour mapping of Arctic Basin ice drift and roughness parameters. J. Geophys. Res., Vol.97, N11, p.17715–17728.

Bradley R.S., Keimig F.T., Diaz H.F. (1992). Climatology of surface-based inversions in the North American Arctic. J. Geophys. Res., Vol.97, ND14, P.15699–115712.

Bradley R.S., Keimig F. T. (1993). Recent changes in the North American Arctic boundary layer in winter. J. Geophys. Res., Vol.98, ND5, p.8851–8858.

Braithwaite R.J. (1993). Is the Greenland ice sheet getting thicker? Guest Editorial. Climatic Change, Vol.23, N4, p.379–382.

Brand A.A. (1963). Study of dielectric properties in microwave region. Moscow, Gosizdat. 403 pp. (in Russian).

Brekhovskikh L.M. (1957). Waves in stratified media. Moscow, AN SSSR publ. 511 pp. (in Russian).

Brekhovskikh L.M. (1978). On the techniques to calculate the thermal regime of emission of randomly inhomogeneous stratified clouds. Izv. AN SSSR, FAO, Vol.XIV, N9, p.997–998 (in Russian).

Bridgman H.A., Schnell R.C., Kahl J.D., Herbert G.A., Joranger E. (1989a). A major haze event near Point Barrow, Alaska: analysis of probable source regions and transport pathways. Atmos. Environm., Vol.23, N11, p.2537–2550.

Bridgman H.A., Schnell R.C., Herbert G.A., Bodhaine B.A., Oltmans S.J. (1989b).

Meteorology and haze structure during AGASP-II. Part II: Canadian arctic flights, 13–16 April, 1986. Atmos. Chem., Vol.9, N1–3, p.49–70.

Brock C.A., Radke L.F., Lyons J.H., Hobbs P.V. (1989). Arctic hazes and the North American Arctic. I: Incidence and origins. J. Atmos. Chem., Vol.9, N1–3, p.129–148.

Bromwich D.H., Robasky F.M. (1993). Recent precipitation trends over the polar ice sheets. Meteorol. and Atmos. Phys., Vol.51, N3–4, p.259–273.

Bromwich D.H., Tzeng R.-Y., Parish T.R. (1994). Simulation of the modern Arctic climate by the NCAR CCM 1. J. Climate (in press).

Brooks K. (1952). Climates of the past. Moscow, Foreign Liter. Publ. 357 pp. (in Russian).

Browell E.V., Butler C.F., Kooi S.A., Fenn M.A., Harriss R.C., Gregory G.L. (1992). Large-scale variability of ozone and aerosols in the summertime arctic and subarctic troposphere. J. Geophys. Res., Vol.97, ND15, p.16433–16450.

Budd W.R. (1991). Antarctica and global change. Clim. Change, Vol.18, N2–3, p.271–300.

Bukharov M.V., Nikitin P.A., Golovnia V.A. (1990). Special assessment of the state of sea ice from radar and radiometric images from the oceanographic satellites. Trudy GosNITsIPR, Issue 37, p.175–185 (in Russian).

Caldeira K. and Kasting J.F. (1994). Global warming on the margin. Nature (in press).

Campbell W.J., Weeks W.F., Ramseier R.O., Gloersen P. (1975). Geophysical studies of floating ice by remote sensing. J. Glaciol., Vol.15, N73, p.305–328.

Campbell W.J., Ramseier R.O., Weeks W.F., Gloersen P. (1976). An integrated approach to the remote sensing of floating ice. Proc. XXVI IAF Congress. Pergamon Press, Oxford. p.445–487.

Campbell W.J. et al. (1978). Microwave remote sensing of sea ice in the IJEX main experiment. Boundary-Layer Meteorol., Vol.13, p.309–337.

Campbell W.J., Gloersen P., Zwally H.J. (1994). Short- and long-term temporal behaviour of polar sea ice cover from satellite passive-microwave observations. Geophysical Monograph 85. AGU, Washington D.C. p.505–520.

Carsey F. (1992). Remote sensing of ice and snow: review and status. Int. J. Remote Sens., Vol.13, N1, p.5–12.

Carsey F.D., Barry R.G., Rothrock D.A., Weeks W.F. (1992). Status and future directions for sea ice remote sensing. In Microwave Remote Sensing of Sea Ice, AGU Monograph 68, edited by F. Carsey. American Geophysical Union, Washington D.C. p.443–446.

Carswell A.I., Ulitcky A., Wardle D.I. 91994). Lidar measurements of the Arctic stratosphere. Preprint. 15 pp.

Cattle H., Murphy J.M., Senior C.A. (1992). The response of Antarctic climate general circulation model experiments with transiently increasing carbon dioxide concentrations. Phil. Trans. Roy. Soc. London B, Vol.338, N12851, p.209–218.

Cavalieri D.J., Gloersen P., Campbell W.J. (1984). Determination of sea-ice parameters with the Nimbus-7 SMMR. J. Geophys. Res., Vol.89, p.5355–5369.

Chang A.T.C., Foster J.L., Hall D.K. (1987). Nimbus-7 SMMR derived global snow cover parameters. Annals of Glaciology, Vol.9, p.39–44.

Chang A.T.C., Foster J.L., and Hall D.K. (1990). Satellite estimates of Northern Hemisphere snow volume, Rem. Sens. Lett., Vol.11, p.167–172.

Chanin M.-L. (Ed.) (1993). The Role of the stratosphere in Global Change. NATO ASI Series, Subseries I "Global Environmental Change", Vol.8, 557 pp.

Chapman W.L., Walsh J.E. (1993). Recent variations of sea ice and air temperatures in high latitudes. Bulletin American Meteorological Society, Vol.74, p.33–47.

Charles C.D., Rind D., Jouzel J., Koster R.D., Fairbanks R.G. (1994). Glacial–interglacial changes in moisture sources for Greenland: influences on the ice record of climate. Science, Vol.263, N5146, p.508–511.

Cheng A., Preller R. (1992). An ice–ocean coupled model for the Northern Hemisphere. Geophys. Res. Lett., Vol.19. N9, p.901–904.

Clarke A.D. (1989). In-situ measurements of the aerosol size distributions, physicochemistry and light absorption properties of arctic haze. J. Atmos. Chem., Vol.9, N1–3, p.255–266.

Clarke A.D., Noone K.J. (1985). Soot in the Arctic Snowpack—a case for perturbations in radiative transfer. Atmos. Environ., Vol.19, N12, p.2045–2053.

Claud C., Scott N.A., Chedin A. (1992). Use of TOVS observations for the study of polar and arctic lows. Int. J. Remote Sens., Vol.13, N1, p.129–140.

Climate Change. The IPCC Scientific Assessment (1990). Ed. by J.T. Houghton, G.J. Jenkins, J.S. Ephraums. Cambridge University Press, Cambridge. 365 pp.

Cohen J. (1994). Snow cover and climate. Weather, Vol.49, N5, p.150–156.

Cohen J., Rind D. (1991). The effect of snow cover on climate. J. Climate, Vol.7, p.698–706.

Colman R.A., McAveney B.J., Fraser J.R., Rikus L.J., Dahni R.R. (1994), Snow and cloud feedbacks modelled by an atmospheric general circulation model. Climate Dynamics, Vol.9, N4/5, p.253–265.

Comiso J.C. (1986). Characteristics of Arctic winter sea ice from satellite multispectral microwave observations. J. Geophys. Res., Vol.91, NC8, p.975–994.

Comiso J.C. (1994). Surface temperatures in the polar regions from Nimbus 7 temperature humidity infrared radiometer. J. Geophys. Res., Vol.99, NC3, p.5181–5200.

Connolley W.M., King J.C. (1993). Atmospheric water vapour transport to Antarctic inferred from radiosonde data. Quart. J. Roy. Meteorol. Soc., Vol.119, Part B, N510, p.325–342.

Covey C., Taylor K.E., Dickinson R.E. (1991). Upper limit for sea ice albedo feedback contribution to global warming. J. Geophys. Res., Vol.96, ND5, p.9169–9174.

Cox P.M., Christensen T., Warner C. (1993). Permafrost. The Hadly Centre for Climate Prediction and Research. Progress Report 1990–1992 and future programme of research. U.K. Met. Office, Bracknell. 58 pp.

Cracknell A.P. (Ed.) (1981). Remote Sensing in Meteorology, Oceanography and Hydrology. Ellis Horwood, Chichester. 542 pp.

Cracknell A., Hayes L. (1988). Introduction to Remote Sensing. Taylor & Francis, Basingstoke. 220 pp.

Craig G.C. (1995). Radiation and polar lows. Quart. J. Roy. Meteorol. Soc., Vol.121, Part A, N521, p.79–94.

Crane R.G., Walsh J.E. (1991). A comparison of global climate model simulations of polar regions. Fifth Conf. on Climate Variations. Amer. Meteorol. Soc., Boston. p.488–491.

Cumming W. (1952). The dielectric properties of ice and snow at 3.2 cm. J. Appl. Phys., Vol.23, p.768–772.

Curry J.A., Ebert E.E. (1989). Sensitivity of the thickness of arctic sea ice to the optical properties of clouds. Glaciology, Vol.17, p.38–51.

Curry J.A., Ebert E.E. (1992). Annual cycle of radiative fluxes over the Arctic Ocean: sensitivity to cloud optical properties. J. Climate, Vol.5, p.1267–1280.

Curry J.A., Schramm J.L., Ebert E.E. (1993). Impact of clouds on the surface radiation budget of the Arctic Ocean. Meteor. and Atmos. Phys., Vol.57, p.197–217.

Darovskikh A.N. (1984). Study of microwave emission of arctic sea ice as applied to the remote sensing diagnostics problem. Leningrad, IAA. Master's thesis (in Russian).

Debye P. (1957). Polar molecules. New York, Dover. 90 pp.

Debye P., Suck G. (1936). Theory of the electric properties of molecules. Moscow– Leningrad. 144 pp. (in Russian).

Deshler T., Adriani A., Gobby G.P., Hofmann D.J., Di Donfrancesco G., Johnson B.J. (1992). Volcanic aerosol and ozone depletion within the Antarctic polar vortex during the austral spring of 1991. Geophys. Res. Lett., Vol.19, p.1819–1822.

DeRycke R.J. (1973). Sea ice motions off Antarctica in the vicinity of the eastern Ross Sea as observed by satellite. J. Geophys. Res., Vol.78, pp.8873–8879.

Dey B. (1981). Monitoring winter sea ice dynamics in the Canadian Arctic with NOAA- TIR images. J. Geophys. Res., Vol.86, p.3223–3235.

Dey B., Felman U. (1989). Observations of winter polynyas and fractures using NOAA AVHRR TIR images and Nimbus-7 SMMR sea ice concentration charts. Remote Sens. Environ., Vol.30, N2, p.141–150.

Dey B., Moore H., Gregory A.F. (1979). Monitoring and mapping sea-ice breakup and freezeup of Arctic Canada from satellite imagery. Arctic Alpine Res., Vol.11, p.229– 242.

Dewey K.F. (1987) Satellite-derived maps of snow cover frequency for the Northern hemisphere. J. Climat. Appl. Meteorol., Vol.26, N9, p.1210–1229.

Doskey P.V., Gaffney J.S. (1992). Non-methane hydrocarbons in the arctic atmosphere at Barrow, Alaska. Geophys. Res. Lett., Vol.19, N4, p.381–384.

Draft plan for the Global Climate Observing System (GCOS) (1993). WMO, Geneva. 36 pp.

Drake F. (1993). Global cloud cover and cloud water path from ISCCP C2 data. Int. J. Climatol., Vol.13, N6, p.581–606.

Dynamics of the Arctic Ocean. A Research Strategy for 90's (1990). Nansen Environ. and Remote Sens. Center, Bergen. 20 pp.

Eder F.X. (1947). Das elektrische Verhalten von Erde (Anomale Dispersion und Absorp- tion). Annalen der Physik, Vol.1, N78, 6 pp.

Emery C.A., Haberle R.M., Ackerman T.P. (1992). A one-dimensional modelling study of carbonaceous haze effects on the springtime arctic environment. J. Geophys. Res., Vol.97, ND18, p.20599–20614.

England A.W. (1979). Thermal microwave emission from a half-space containing scatterers. Radio Science, N9, p.447–454.

England A.W., Johnson K. (1975). Thermal microwave detection of near-surface thermal anomalies. Proc. of U.N. Geothermal Symposium, Frozen Grounds. p.971–979.

Eppler D.T., Farmer L.D., Lohanick A.W., Anderson M.R., Cavalieri D.J., Comiso J., Gloersen P., Garrity C., Grenfell T.C., Hallikainen M., Maslanik J.A., Mätzler C., Melloh R.A., Rubenstein I., Swift C.T., (1992). Passive microwave signatures of sea ice. In Microwave Remote Sensing of Sea Ice, AGU Monograph 68, edited by F. Carsey, American Geophysical Union, Washington D.C. p.47–72.

Evans S. (1965). Dielectric properties of ice and snow. J. Glaciol., Vol.2, N42, p.773–792.

Fan S.M., Wofsy S.C., Bakwin P.S., Jacob D.J., Anderson S.M., Kebabian P.L., McManus J.B., Kolb C.E., Fitzjarrald D.R. (1992). Micrometeorological measurements of CH_4 and CO_2 exchange between the atmosphere and subarctic tundra. J. Geophys. Res., Vol.97, ND15, p.16627–16644.

Federal Arctic Research (1985). Detailed listing of existing U.S. programs. Integral Compilation. September 1985. Prepared by The Interagency Arctic Research Policy Committee. DOE/ER-0251. Washington D.C. 136 pp.

First Global GARP Experiment (1981). Vol.2, Polar aerosol, extended cloudiness and radiation. Ed. by K.Ya. Kondratyev, V.F. Zhvalev. Leningrad, Gidrometeoizdat. 150 pp. (in Russian).

Fisher A.D., Ledsham B.L., Rosenkranz P.W., Staelin D.H. (1976). Satellite observations of snow and ice with an imaging passive microwave spectrometer. COSPAR Proc. of Symp. on Meteorol. Observations from Space: Their Contribution to the FGGE. Philadelphia. p.98–103.

Foster J.L. (1989). The significance of the date of snow disappearance on the arctic tundra as a possible indicator of climate change. Arct. and Alp. Res., Vol.21, N1, p.60–70.

Foster J.L., Chang A.T.C. (1993). Snow cover. In Gumey et al., Atlas of Satellite Observations Related to Global Change. Cambridge University Press, Cambridge. p. 361–370.

Frederick J.E., Alberts A.D. (1991). Prolonged enhancement in surface ultraviolet radiation during the Antarctic spring of 1990. Geophys. Res. Lett., Vol.18, N10, p.1869–1872.

Frederick K.D. (1994). Integrated assessments of the impacts of climate change on natural resources. An introductory essay. Clim. Change, Vol.28, N1–2, p.1–14.

Fröhlich G. (1960). Theory of dielectrics. Moscow, Foreign Liter. Publ. (in Russian).

Garcia R.R., Boville B.A. (1994). "Downward control" of the mean meridional circulation and temperature distribution of the polar winter stratosphere. J. Atmos. Sci., Vol.51, N15, p.2238–2245.

Gates W.L. (1992). AMIP: The Atmospheric Model Intercomparison Project. Bull. Amer. Met. Soc., Vol.73, p.1962–1970.

Gavrilenko A.S., Kryzhanovsky V.V., Kuleshov Yu.A., et al. (1986). A radiophysical complex to remotely sound the environment. Preprint no. 321. Ukr. SSR Acad. Sci., Inst. Radiophys. and Electronics, Kharkov. 40 pp. (in Russian).

Genthon C. (1994). Antarctic climate modeling with general circulation models of the atmosphere. J. Geophys. Res. D, Vol.99, N6, p.12953–12961.

Gernet E.S. (1930). Ice lichens. Tokyo. 122 pp. (New edition: Moscow, Nauka Publ. 144 pp. (in Russian).)

Glaciers, ice sheets, and sea level: effect of a CO_2-induced climatic change (1985). Report of a workshop held in Seattle, Washington. September 13–15, 1984. U.S. Dept. of Energy, DOE/ER/602351, Washington D.C. 330 pp.

Gleick P.H., Rango A. (1994). Evaluating climate change impacts in snowmelt basins. Eos, Vol.75, N09, p.107.

Glendening J.W. (1992). Vertical distribution of the heat from arctic leads and its corporation into large-scale models. Proc. Third Symp. on Global Change Studies. Amer. Meteorol. Soc., Boston. p.60–62.

Global Climate Modelling (1992). Report of second session of WCRP Steering Group on Global Climate Modelling (Bristol, U.K., 18–20 November 1991). WCRP-7 (WMO/TD-N 482). Geneva. 37 pp.

Gloersen P., Campbell W.J. (1988). Variations in the Arctic, Antarctic, and global sea ice covers during 1978–1987 as observed with the Nimbus-7 SMMR. J. Geophys. Res., Vol.693, N8, p.10666–10674.

Gloersen P., Campbell W.J. (1991). Recent variations in Arctic and Antarctic sea-ice covers. Nature, Vol.352, p.33–36.

Gloersen P., Nordberg W., Schmuge T.J. et al. (1973). Microwave signatures of first-year and multi-year sea ice. J. Geophys. Res., Vol.78, N18, p.3564–3572.

Gloersen P., Ramseier R.O., Campbell W.J., Chen T.C., Wilheit T.T. (1975a). Variation of ice morphology of selected mesoscale test areas during the Bering Sea Experiment. Soviet–American Bering Sea Experiment. Proc. Final Symp. on results of the joint Soviet–American expedition. Leningrad, Gidrometeoizdat. P.234–270 (in Russian).

Gloersen P., Wilheit T.T., Chang T.C., Nordberg W., Campbell W.J. (1975b). Microwave maps of the polar ice of the Earth. Climate Arct. Fairbanks, p.407–414.

Gloersen P., Campbell W.J., Cavalieri D.J., Comiso J.C., Parkinson C.L., Zwally H.L. (1992). Arctic and Antarctic Sea Ice, 1978–1987: Satellite Passive Microwave Observations and Analysis, NASA SP-511.

Glowienska-Hense R., Hense A. (1992). The effect of an arctic polynya on the northern hemisphere mean circulation and eddy regime: a numerical experiment. Climate Dynamics, Vol.7, N3, p.155–163.

Gobby G.P., Adriani A. (1993). Mechanisms of formation of stratospheric clouds observed during the Antarctic late winter of 1992. Geophys. Res. Lett., Vol.20, N14, p.1427–1430.

Goody R. (1980). Polar processes and world climate (a brief review). Mon. Weather. Rev., Vol.108, N12, p.1935–1942.

Goody R.M., Yung Y.L. (1989). Atmospheric Radiation. Theoretical Basis. Oxford University Press, New York. 519 pp.

Gorsky A.I. (1957). Some measurements at centimeter wavelengths using the geometric optics technique. Instruments and Experimental Technology, N1, p.87–89 (in Russian).

Graf H.-F. (1992). Arctic radiation deficit and climate variability. Climate Dynam. Vol.7, N1, p.19–28.

Grenfell T.C., Warren S.G., Mullen P.C. (1994). Reflection of solar radiation by the Antarctic snow surface at ultraviolet, visible and near-infrared. J. Geophys. Res., Vol.99, ND9, p.18669–18684.

Gurney R.J., Foster J., Parkinson C. (Eds) (1993). Atlas of Satellite Observations. Cambridge University Press. 350 pp.

Gurvich A.S., Kalinin Ts.I., Matveev D.T. (1973). The effect of internal structure of glaciers on their thermal microwave emission. Izv. AN SSSR, FAO, Vol.IX, N12, p.1247–1255 (In Russian).

Gvishiani J. (1994). From the decline to the growth, from "The Limits to Growth" to the growth of selfconscience. Russian Province, N5, p.98–102 (in Russian).

Hakkinen S. (1993). An Arctic source for the Great Salinity Anomaly: a simulation of the Arctic ice-ocean system for 1955–1975. J. Geophys. Res., Vol.98, NC9, p.16397–116410.

Hakkinen S., Mellor G.L. (1992). Modeling the seasonal variability of a coupled Arctic ice-ocean system. J. Geophys. Res., Vol.97, NC12, p.20285–20304.

Hall D.K. (1988). Assessment of polar climatic change using satellite technology. Rev. Geophys., Vol.6, N21, p.26–39.

Hall D.K., Chang A.T.C., Foster J.L. (1986). Detection of the depth-hoar layer in the snowpack of the Arctic coastal plain of Alaska, USA, using satellite data. J. Glaciol., Vol.32, p.87–94.

Hall D.K., Martinec J. (1986). Remote Sensing of Ice and Snow. Chapman & Hall, London. 189 pp.

Hallikainen M. (1978) Measured permittivities of snow and low-salinity sea ice for UHF radiometer applications. Presented at Swedish Nat. Convention Radio Sci., Stockholm, Sweden, March 29–31.

Hallikainen M., Ulaby F.T., Abdelrazik M. (1986). Dielectric properties of snow in the 3 to 37 GHz range. IEEE Transact. on Antennas and Prop., Vol.AP-34, N11, p.1329–1340.

Hansen A.D.A., Bodhaine B.A., Dutton E.G., Schnell R.C. (1988) Aerosol black carbon measurements at the South Pole: initial result. Geophys. Res. Lett., Vol.15, N11, p.1193–1196.

Hansen A.D.A., Conway T.J., Steele L.P., Bodhaine B.A., Thoning K.W., Tans P., Novakov Y. (1989). Correlations among combustion effluent species at Barrow, Alaska: aerosol black carbon, carbon dioxide, and methane. J. Atmos. Chem., Vol.9, N1–3, p.283–300.

Harries J.E. (1990). Earthwatch: the Climate from Space. Ellis Horwood, Chichester. 150 pp.

Harries J.E., Russel J.M., III, Park J., Tuck A.F., Drayson S.R. (1995). Observations of absorbing layers in the Antarctic stratosphere in October 1991. Quart. J. Roy. Meteorol. Soc., Vol.121 (in press).

Harris C., Stonehouse B. (1991). Antarctica and Global Climate Change. Belhaven Press, London. 215 pp.

Harriss R.C., Sachse G.W., Hill G.F., Wade L., Bartlett K.B., Collins J.E., Steele P., Novelli P. (1992). Carbon monoxide and methane in the North American Arctic and subarctic troposphere: July–August 1988. J. Geophys. Res., Vol.97, ND15, p.16589–16600.

Harte J., Williams J. (1988). Arctic aerosol and arctic climate: results from an energy budget model. Climate Change, Vol.13, N2, p.161–189.

Hartman D.L. (1994). Global Climatology. Academic Press, New York. 397 pp.

Harvey L.D.D. (1992). A two-dimensional ocean model for long-term climatic simulations: stability and coupling to atmospheric and sea ice models. J. Geophys. Res., Vol.970, N6, p.9435–9454.

Hempel (Ed.) (1994). Antarctic Science: Global Concerns. Springer, Heidelberg. 287 pp.

Herman G., Goody R. (1976). Formation and persistence of summertime arctic status clouds. J. Atmos. Sci., Vol.33, p.1537–1553.

Hibler W.D. III (1992). The role of sea ice dynamics in global climate change. In Modeling the Earth System (Ed. by D. Ojima), Vol.3. UCAR OIES. Global Change Insdt., Boulder, CO. p.107–130.

Hibler W.D. III, Thorndike A.S. (1992). Report: Cryosphere and Climate. In Modeling the Earth System (Ed. by T. Ojima), Vol.3. UCAR OIES. Global Change Inst., Boulder, CO. p.327–334.

Hines M.E., Morrison M.C. (1992) Emissions of biogenic sulfur gases from Alaskan tundra. J. Geophys. Res., Vol.97, ND15, p.16703–16708.

Hinzman L.D., Kane D.L. (1992). Potential response of an Arctic watershed during a period of global warming. J. Geophys. Res., Vol.97, ND3, p.2811–2820.

Hippel A.R. (1959). Dielectrics and their application. Moscow, Gosenergoizdat (in Russian).

Hobbs P.V. (Ed.) (1993). Aerosol Cloud–Climate Interactions. Academic Press, New York. 222 pp.

Hofmann D.J., Oltmans S.J., Deshler T. (1991). Simultaneous balloonborne measurements of stratospheric water vapour and ozone in the polar regions. Geophys. Res. Lett., Vol.18, N6, p.1011–1014.

Hofmann D.J., Ferguson E.E., Johnson P.V., Matthews W.A. (1992). Tropospheric ozone variations in the Arctic during January 1990. Planet. Space Sci., Vol.40, N2–3, p.203–210.

Hoekstra P. (1970). The dielectric properties of sea ice at UHF and microwave frequencies. Proc. Int. Meet. on Radioglaciology. Lyngby. p.32–53.

Hofer R., Good W. (1979). Snow parameter determination by multichannel microwave radiometry. Remote Sens. Environm., Vol.8, N3, p.211–224.

Houghton John T. (Ed.) (1984). The Global Climate. Cambridge University Press, Cambridge. 233 pp.

Huybrechts P. (1990). A 3-D model for the Antarctic ice sheet: a sensitivity study on the glacial–interglacial contrast. Climate Dynam., Vol.5, N2, p.79–92.

Huybrechts P., Oerlemans J. (1990). Response of the Antarctic ice sheet to future greenhouse warming. Climate Dynam., Vol.5, N2, p.93–102.

Identifying properties for investment in developing new technology for the ECOPS Grand Challenger. Report of a Workshop held at Brighton, U.K., 6–8 March 1994. Commission of the European Communities. 16 pp.

International Conference on the Role of the Polar Ocean in Global Change (1991). Fairbanks, Alaska.

International Conference for Arctic Research Planning, 5–9 December 1995 in Hanover, New Hampshire, USA. 1994 IASC Progress, N3, p.1.

Ito H., Muller F. (1982). Ice movement through Smith Sound in northern Baffin Bay, Canada, observed in satellite imagery. J. Glaciol. Vol.9, p.92–96.

IUGG XX General Assembly (1991). Vienna, August. Program and Abstract. p.89–90.

Iversen T. (1989). Numerical modelling of the long range atmospheric transport of sulphur dioxide and particulate sulphate to the Arctic. Atm. Environ., Vol.23, N11, p.2571–2596.

Iwaska Y., Hayashi M., Kondo Y., Koike M., Koga S., Yamato M., Aimedieu P., Matthews W.A. (1992). Chemical state of polar stratospheric aerosols. Proc. NIPR Symp. Polar Meteorol. and Glaciol., N5, p.1–8.

Jacob D.J., Fan S.-M., Wofsy S.C., Spiro P.A., Bakwin P.S., Ritter J.A., Browell E.V., Gregory G.L., Fitzjarrald D.R., Moore K.E. (1992). Deposition of ozone to tundra. J. Geophys. Res., Vol.97, ND15, p.16473–16480.

Jaworowski Z. (1989). Pollution of the Norwegian Arctic: a review. Rapportserie No. 55. Norsk Polarinstitutt. Oslo. 93 pp.

Jaworowski Z. (1993). Fallout studies show that marine chlorine reaches stratosphere. Res. Commun. 21st Century. p.6–7.

Jaworowski Z. (1994). Ancient atmosphere-validity of ice records. Environ. Sci. & Pollut. Res., Vol.1, N3, p.161–171.

Jaworowski Z., Segalstad T.V., Hisdal V. (1990). Atmospheric CO_2 and global warming: a critical review. Norsk Polarinstitutt, Rapportserie No. 39, Oslo. 75 pp.

Jaworowski Z., Segalstad T.V., Oho N. (1992). Do glaciers tell a true atmospheric CO_2 story? The Sea Tot. Environ., Vol.114, p.227–284.

Jennings S.G. (Ed.) (1993). Aerosol Effects on Climate. The University of Arizona Press. 305 pp.

Johannessen O.M. (1987) Introduction: Summer Marginal Ice Zone Experiment during 1983 and 1984 in Fram Strait and the Greenland Sea. J. Geophys. Res., Vol.92, p.6716–6717.

Johannessen J.A., Roed L.P., Johannessen O.M., Evensen G., Hackett B., Pettersson L.H., Haugan P.M., Sandven S., Shuchman R. (1992). Monitoring and modelling of the marine coastal environment. Preprints. First Thematic Conf. on Remote Sensing for Marine and Coastal Environm, New Orleans. 18 pp.

Johannessen O.M., Miles, M., Bjørgo E. (1994). Sea ice concentration derived from SMMR and SSMI: time series analysis for climate change detection. Proceedings European Symposium on Remote Sensing, 26–30 September, 1994, Rome.

Johannessen O.M., Miles M, Bjørgo E. (1995). Trends in Arctic and Antarctic sea ice 1978–1994. Nature (in press).

Johannessen O.M., Campbell W.J., Shuchman R., Sandven S., Gloersen P., Josberger E.G., Johannessen J.A., Haugan P.M. Seasonal ice zone processes in the Greenland

and Barents Sea from microwave observations. Microwave Remote Sensing of Sea Ice (in press).

Johns E.B. (1983). Snowpack ground-truth manual. NASA Contract Report 170584.

Kahl J.D., Hansen T.J. (1989). Determination of regional sources of aerosol black carbon in the Arctic. Geophys. Res. Lett., Vol.16, N4, p.327–330.

Kahl J.D., Charlevoix D.J., Zaitseva N.A., Schnell R.C., Serreze M.C. (1993a). Letters to Nature. Vol.361, p.335–337.

Kahl J.D.W., Serreze M.C., Stone R.S., Shiotani S., Kisley M., Schnell R.C. (1993b). Tropospheric temperature trends in the Arctic: 1958–1986. J. Geophys. Res., Vol.98, ND7, p.12825–12838.

Kalmykov A.I., Efimov V.B., Kavelin S.S. et al. (1984). The "Kosmos-1500" radar system. Study of the Earth from Space, N5, p.84–93 (in Russian).

Katsov V.M., Meleshko V.P., Sokolov A.P., Lyubanskaya V.A. (1993). Impact of sea ice on thermal regime and atmospheric circulation during winter of the northern hemisphere. Meteorol. and Hydrol., N12, p.5–24.

Khokhlov G.P. (1978). Physico-chemical properties of the upper ice layers of different age in the NP-22 region. Proc. IAA, Vol.359, p.4–12 (in Russian).

Kondratyev K.Ya. (1986). Changes in Global Climate. A.A. Malkema, Rotterdam, 280 pp.

Kondratyev K.Ya. (1988). Climate Shocks: Natural and Anthropogenic. John Wiley, New York. 296 pp.

Kondratyev K.Ya. (1989). Global Ozone Dynamics. Progress in Science and Technology. Geomagnetism and upper atmosphere, Vol.11. Moscow, VINITI. 212 pp. (in Russian).

Kondratyev K.Ya. (1990). International Space Year: priorities and perspectives. Studies of the Earth from Space. N1, p.3–15 (in Russian).

Kondratyev K.Ya. (1991a). New assessments of global climate change. Atmosphera, Vol.4, N3, p.177–188.

Kondratyev K.Ya. (1991b). On the International Framework Convention on climate change. Izv. AN SSSR. Ser. Geogr., N6, p.117–122 (in Russian).

Kondratyev K.Ya. (1992a). Global Climate. Leningrad, Nauka Publ. 359 pp. (in Russian).

Kondratyev K.Ya. (1992b). On the Framework Convention on Climate Change. Il Nuovo Cim. C, Vol.156, N1, p.87–98.

Kondratyev K.Ya. (1992c). Aerosol–cloud–climate interaction. Pt. 1: Aerosols. P. 2: Clouds. Optics of the Atmosphere, Vol.5, N3, p.317–323, 324–325 (in Russian).

Kondratyev K.Ya. (1995). European Programme of studies of the polar regions and the ocean. Izv. Russian Geographical Soc., Vol.127, N1, p.38–42 (in Russian).

Kondratyev K.Ya., Binenko V.I. (1981). Atmospheric aerosol of polar regions. First Global GARP Experiment. Vol.2. Polar aerosol, extended cloudiness and radiation. Leningrad, Gidrometeoizdat. p.92–100 (in Russian).

Kondratyev K.Ya., Cracknell A.P. (1995). Observing Global Climate Change. Taylor & Francis, London (in press).

Kondratyev K.Ya., Kotliakov V.M. (1991). Polar regions and their contribution to the formation of global ecological changes. Izv. AN SSSR, Ser. Geogr., N1, p.36–44 (in Russian).

Kondratyev K.Ya., Johannessen O.M. (1993). Arctic and Climate. PROPO, St. Petersburg. 144 pp. (in Russian).

Kondratyev K.Ya., Melentyev V.V. (1994). Microwave remote sensing of the snow and ice cover: the Russian experience. Geophysical Monograph 85, AGU, Washington D.C. p.497–504.

Kondratyev K.Ya., Zhvalev V.F. (eds.) (1981). First Global GARP Experiment. Vol.2. Polar aerosol, extended cloudiness and radiation. Leningrad, Gidrometeoizdat. 150 pp. (in Russian).

Kondratyev K.Ya., Melentyev V.V., Rabinovich Yu.I. (1973a). The Soviet–American Bering Sea Experiment. Meteorology and Hydrology, N11, p.3–10 (in Russian).

Kondratyev K.Ya., Melentyev V.V., Rabinovich Yu.I., Shulgina E.M. (1973b). Microwave sounding of ice cover. Proc. Symp. on Physical Methods to Study Snow and Ice. Leningrad, Gidrometeoizdat. p.15–16 (in Russian).

Kondratyev K.Ya., Rabinovich Yu.I., Melentyev V.V. (1973c). Soviet–American Experiment "Bering". Proc. Symp. on Physical Methods to Study Snow and Ice. Leningrad, Gidrometeoizdat. p.3–15 (in Russian).

Kondratyev K.Ya., Vlasov V.P., Melentyev V.V. (1985). Radiothermal emission of the melting ice cover as an indicator of its state (with lake Sevan as an example). Doklady AN SSSR, Vol.280, N4, p.839–842 (in Russian).

Kondratyev K.Ya., Vlasov V.P., Kalmykov A.I. (1987). A study of the lake Onega ice cover using the side-looking radar. Third Symp. of the Soviet Oceanologists: Summary of reports. Section of Physics and Chemistry of the Ocean (climate, ocean–atmosphere interaction, space oceanology). Leningrad, Gidrometeoizdat. p.145–146 (in Russian).

Kondratyev K.Ya., Dyachenko L.N., Kozoderov V.V. (1988a). Earth Radiation Budget. Leningrad, Gidrometeoizdat. 350 pp. (in Russian).

Kondratyev K.Ya., Kalmykov A.I., Vlasov V.P., Kuleshov Yu.A., Timchenko A.I. (1988b). Complex radar measurements of the freshwater ice cover. Doklady AN SSSR, Vol.298, N2, p.317–320 (in Russian).

Kondratyev K.Ya., Melentyev V.V., Alexandrov V.Yu. (1989). Microwave emittance properties of different types of the surface for negative temperatures. Doklady AN SSSR, Vol.306, N1, p.67–70 (in Russian).

Kondratyev K.Ya., Loginov V.F., Kravchuk E.G. (1990). Solar radiation trends as indicators of changes due to anthropogenic aerosols. Doklady USSR Acad. Sci., Vol.315, N2, p.341–344 (in Russian).

Kondratyev K.Ya., Efremov G.A., Buznikov A.A., Viter V.V., Gorozov V.A., Chekhin L.P. (1991). Experimental processing of high-resolution radar images using the synthetic-aperture system. Doklady AN SSSR, Vol.317, N1, p.70–77 (in Russian).

Kondratyev K.Ya., Bondarenko V.G., Khvorostyanov V.I. (1992a). A three-dimensional numerical model of cloud formation and aerosol transport in the orographically inhomogeneous atmospheric boundary layer. Boundary-Layer Meteorol., Vol.11, N5, p.265–285.

Kondratyev K.Ya., Melentyev V.V., Nazarkin B.A. (1992b). The spaceborne remote sensing of water basins and catchment areas. SPb, Gidrometeoizdat. 248 pp. (in Russian).

Koster R.D., Suarez M.J. (1993). Climate variability studies with a coupled land/atmosphere model. IAHS Publ. No. 214. p.173–181.

Kotliakov V.M. (1968). Earth's snow cover and glaciers. Leningrad, Gidrometeoizdat. 479 pp. (in Russian).

Kotliakov V.M., Krenke A.N. (1982). The role of snow cover and glaciers in global climate models. Izv. AN SSSR., Ser. Geogr., N1, p.5–14 (in Russian).

Kottmeier Ch., Engelbart D. (1992). Generation and atmospheric heat exchange of coastal polynyas in the Weddel Sea. Boundary-Layer Meteorol., Vol.6, N3, pp.207–234.

Kullman L. (1992). High latitude environments and environmental change. Prog. Phys. Geog., Vol.16, N4, p.478–488.

Kullman L. (1994). Climate and environmental change at high northern latitudes. Prog. Phys. Geog., Vol.18, N1, p.124–135.

Künzi K.F., Fischer A.D., Staelin D.H., Waters J.W. (1976) Snow and ice surface measured by the Nimbus-5 microwave spectrometer. Geophys. Res., Vol.81,N27, p.4965–4976.

Kvenevolden K.A. (1988). Methane hydrates and global climate. Global Biogeochem. Cycles, Vol.2, p.221–230.

Lange M.A. (1993). Forging new partnerships in the North: the Northern Forum and the Barnets Sea Euro-Arctic Council. Arctic Centre News, N3, p.2–4.

Lappo S.S., Gulev S.K., Rozhdestvensky A.E. (1990). Large-scale thermal interaction in the ocean–atmosphere system and the energy-active zones of the World Ocean. Leningrad, Gidrometeoizdat. 236 pp. (in Russian).

Ledley T.S. (1988). A coupled energy balance climate–sea ice model: impact of sea and leads on climate, J. Geophys. Res., Vol.93, ND12, p.15919–15932.

Ledley T.S. (1991). Snow on sea ice: competing effects in shaping climate. J. Geophys. Res., Vol.96, ND9, p.17195–17208.

Ledley T.S. (1993). Variations in snow on sea ice: a mechanism for producing climate variations. J. Geophys. Res., Vol.98, p.10401–10410.

Legenkov A.P. (1992). Deformation of Drifting Ice in the Arctic Ocean. Gidrometeoizdat, St. Petersburg. 104 pp. (in Russian).

Li Z., Leighton H.G. (1991). Scene identification and its effect on cloud radiative forcing in the Arctic. J. Geophys. Res., Vol.96, ND5, p.9175–9188.

Li Shi.-M., Winchester J.M. 91993). Water soluble organic constituents in Arctic aerosols and snow pack. Geophys. Res. Lett., Vol.20, N1, p.45–48.

Lindsay R.M., Rothrock D.A. (1994). Arctic sea ice surface temperature from AVHRR. J. Climate, Vol.7, p.174–183.

Loehle C. (1993). Geological methane as a source for postglacial CO_2 increases: the hydrocarbon pump hypothesis. Geophys. Res. Lett., Vol.20, N4, p.1415–1418.

Loginov V.F. (1992). Causes and Consequences of Climate Changes. 'Science and Technology', Minsk. 320 pp.

Lohmann U., Roeckner E. (1990). The influence of cirrus cloud-radiative forcing on climate and climate sensitivity in a general circulation model. Max-Planck-Inst. f. Meteorologie Rept. no. 126 p.1–48.

Lubin D. (1994). Infrared radiative properties of the maritime antarctic atmosphere. J. Climate, Vol.7, N1, p.121–140.

MacCracken M.C., Cess R.D., Potter G.L. (1986). Climatic effects of anthropogenic arctic aerosols: an illustration of climate feed-back mechanisms with one- and two-dimensional climate models. J. Geophys. Res., Vol.91, p.14445–14450.

Manabe S., Spelman M.J., Bryan K. (1991). Transient response of a coupled ocean–atmosphere model to gradual changes of atmospheric CO_2. Part II: Annual mean response. J. Climate, Vol.4, p.785–818.

Manabe S., Spelman M.J., Stouffer R.J. (1992). Transient response of a coupled ocean–atmosphere model to gradual changes of atmospheric CO_2. Part II: Seasonal response. J. Climate, Vol.5, p.105–126.

Manabe S., Wetherald R.T. (1980). On the distribution of climate change resulting from an increase of CO_2 content of the atmosphere. J. Atmos. Sci., Vol.37, p.99–118.

Mancini E., Giovani P., Visconti G. (1992). Dehydration on the antarctic stratosphere. Radiative effects. Geophys. Res. Lett., Vol.19, N7, p.585–588.

Man's Future in Arctic Areas (1990). Ed. by R. Kivilahti, C. Kurppa, M. Pretes. Arctic Centre Publications, Rovaniemi. 211 pp.

Marchuk G.I., Kondratyev K.Ya., Kozoderov V.V., Khvorostyanov V.I. (1986). Clouds and Climate. Leningrad, Gidrometeoizdat. 512 pp. (in Russian).

Marchuk G.I., Kondratyev K.Ya., Kozoderov V.V. (1990). Earth Radiation Budget: Key Aspects. Moscow, Nauka Publ. 232 pp.

Martinec J., Rango A., Roberts R. (1994). Modelling the redistribution of runoff caused by global warming. In Effects of Human-Induced Changes of Hydrologic Systems. Amer. Water Resource Assoc. p.153–161.

Martsinkevich L.M., Melentyev V.V. (1972). Emission of the rough sea surface in the centimeter wavelength region. Trudy GGO, Issue 291, P.24–33 (in Russian).

Massie S.T., Bailey P.L., Gille J.C., Lee E.C., Mergenthaler J.L., Roche A.E., Kumer J.B., Fishbein E.F., Waters J.W., Lahoz W.A. (1994). Spectral signatures of polar stratospheric clouds and sulfate aerosol. J. Atmos. Sci., Vol.51, N20, p.3027–3044.

Massom R. (1991). Satellite Remote Sensing of Polar Regions. Belhaven Press, London. 320 pp.

Matson M., Roplewski C.F., Varnadore M.S., (1987). An Atlas of Satellite-Derived Northern Hemisphere Snow Cover Frequency. U.S. Department of Commerce, Washington D.C.

Matveev L.T. (1991). Theory of atmospheric general circulation and Earth's climate. Leningrad, Gidrometeoizdat. 296 pp. (in Russian).

Mätzler C., Schanda E., Good W. (1982). Towards the definition of optimum sensor specifications for microwave remote sensing of snow. IEEE Trans. Geosci. Remote Sens., Vol.20, N1, p.57–61.

McConnel J.C., Henderson G.S., Barrie C., Rottenheim J., Niki H., Langford C.H., Templeton E.M.J. (1992). Photochemical bromine production implicated in Arctic boundary-layer depletion of ozone. Nature. Vol.355, N6356, p.150–152.

McGinnis D.F., Jr, Pritchard J.A., Wiesnet J.R. (1975). Snow depth and snow extent using VHRR data from the NOAA-2 satellite. NOAA Techn. Memo. NESS-63. Washington D.C. 10 pp.

McPhee M.G. (1992). Turbulent heat flux in the upper ocean under sea ice. J. Geophys. Res., Vol.97, p.5365–5379.

Melentyev V.V., Alexandrov V.Yu. (1989). Study of the Sea of Okhotsk ice cover using the airborne microwave scanning radiometer. Electrophysical and Physico-Mechanical Properties of Ice. Leningrad, Gidrometeoizdat. p.78–87 (in Russian).

Melentyev V.V., Alexandrov V.Yu. (1991). Model calculations of microwave emissivity of freshwater ice and frozen soils. Proc. AANII, Vol.421, p.138–146 (in Russian).

Meleshko V.P. (1991). Calculations of anthropogenic climate changes using the coupled global ocean–atmosphere models: problems and ways of further development of models (overview). Izv. AN SSSR, FAO, Vol.27, N7, p.691–723 (in Russian).

Meleshko V.P., Sokolov A.P., Sheinin D.A. et al. (1991). A model of the atmosphere and upper layer of the ocean for climate studies and long-range weather forecasts. Meteorology and Hydrology, N5, p.5–14 (in Russian).

Mellor G.L., Häkkinen S. (1994). A review of coupled ice–ocean models. Geophysical Monograph 85. AGU, Washington D.C. p.21–32.

Mendelsohn R., Rosenberg N.J. (1994). Framework for integrated assessments of global warming impacts. Clim. Change, Vol.28, N1–2, p.15–44.

Meyer F.G., Curry J.A., Brock C.A., Radke L.F. (1991). Springtime visibility in the Arctic. J. Appl. Meteorol, Vol.30, p.342–357.

MGO (1978). Assessment of the ice cover characteristics with microwave remote sensing techniques. MGO Report No. 78032305, Leningrad.

Miles M.W., Barry R.G. (1995). Satellite climatology of leads in arctic sea ice in winter. J. of Geophys. Res. (in press).

Mitnik L.M., Kalmykov A.I. (1992). Structure and dynamics of the Sea of Okhotsk marginal ice zone from "Ocean" satellite radar sensing data. J. Geophys. Res, Vol.97, NC5, p.7429–7446.

MIZEX Group (1989). MIZEX East 1987: The winter Marginal Ice Zone Programme in the Fram Strait/Greenland Sea. EOS, Vol.70, N17, p.545–555.

Moore J.C., Narita H., Maeno N. (1991). A continuous 770-year record of volcanic activity from East Antarctica. J. Geophys. Res., Vol.96, ND9, P.17353–17360.

Morison J.H., McPhee M.G., Curtin T.B., Paulson C.A. (1992). The oceanography of winter leads. J. Geophys. Res., Vol.97, NC7, p.11199–11218.

Morozov P.T., Khokhlov G.P. (1973). Physico-chemical and electric properties of the Bering Sea near-shore ice. Preliminary results from the Bering Sea experiment (preprint). Leningrad. p.53–64 (in Russian).

Morrissey L.A., Livingstone G.P. (1992). Methane emissions from Alaska arctic tundra: an assessment of local spatal variability. J. Geophys. Res., Vol.97, ND15, p.16661–16670.

Mysak L.A., Huang F. (1992). A latent and sensible-heat polynya model for North Water, Northern Baffin Bay. J. Phys. Oceanogr., Vol.22, N6, p.596–608.

Mysak L.A., Manak D.K. (1989). Arctic sea ice extent and anomalies, 1953–84. Atmos. Ocean, Vol.27, N2, p.346–405.

Mysak L.A., Manak D.K., Marsden R.F. (1990). Sea-ice anomalies observed in the Greenland and Labrador Seas during 1901–1984 and their relation to an interdecadal arctic climate cycle. Climate Dynam., Vol.5, N2, p.111–133.

Mysak L.A., Power S.B. (1991). Greenland sea ice and salinity anomalies and interdecadal climate variability. Climate Bull., Vol.25, N2, p.81–91.

Nagurny A.P. (1986). Modelling the continental and sea ice in climate models. Progress in Science and Technology. Meteorology and Climatology, Vol.13. Moscow, VINITI. 104 pp. (in Russian).

Namias J. (1981). Snow covers in climate and long-range forecasting. "Global Data", N11, p.13–26.

Nansen Centennial Arctic Programme (1992). Plan for a scientific expedition 1994–1996. Ed. by T.O. Vorren, O.M. Johannessen et al. Norwegian Research Council for Science and Humanities, Oslo. 53 pp.

NASA (1982). Plan of research for snowpack properties remote sensing (PRS) recommendations of the Snowpack Properties Working Group, Goddard Space Flight Center, Greenbelt, MD.

National Issues and Research Priorities in the Arctic Polar Research Board (1985). Washington D.C. 124 pp.

Netushil A.V. (1975). Models of electric fields in heterogenic media of irregular structures. Electricity, N10, p.1–8 (in Russian).

Newell R.E., Wu Z.-X. (1992). Interrelationship between temperature changes in the free atmosphere and sea surface temperature changes. J. Geophys. Res., Vol.97, ND4, p.3693–3710.

Newell R.E., Wu Z.-X. (1994). Temperature limits in the climate system. Proc. Indian Natn. Sci. Acad., Vol.60, N1, p.1–22.

Newell R.E., Zhu Y. (1994). Tropospheric rivers: a one-year record and a possible application to ice core data. Geophys. Res. Lett., Vol.21, N2, p.113–116.

Niki H., Becker K.H. (Eds) (1993). The Tropospheric Chemistry of Ozone in the Polar Regions. Springer, Heidelberg. 420 pp.

Nikitin P.A. (1980). Variations of microwave sea ice emission (results) from numerical experiment. Study of the Earth from Space, N5, p.56–63 (in Russian).

Nyfors E. (1983). On the dielectric properties of dry snow in the 800 MHz to 13 GHz region. Helsinki University of Technology, Radio Lab., Otaniemi, Finland, Rep. S135.

Oerlemans J., van der Veen (1984). Ice Sheets and Climate. D. Reidel, Dordrecht. 228 pp.

Olhoert G.R. (1978). Electrical properties of permafrost. Proc. 3rd Int. Conf. Permafrost, Edmonton, Vol.1, p.127–132.

Omstedt A., Wettlaufer J.S. (1992). Ice growth and oceanic heat flux. J. Geophys. Res., Vol.97C, N6, p.9383–9390.

Ottar B. (1989). Arctic air pollution: a Norwegian perspective. Atmos. Environm., Vol.23, N11, p.2349–2357.

Parashar S.K., Fung A.K., Moore R.K. (1978). A theory of wave scatter from an inhomogeneous medium with a slightly rough boundary and its application to sea ice. Remote Sens. Environm., Vol.7, N1, p.37–50.

Park J.H., Russel, III, J.M. (1994). Summer polar chemistry observations in the strato sphere made by HALOE. J. Atmos. Sci., Vol.51, N20, p.2903–2913.

Parkinson C.L. (1991). Interannual variability of the spatial distribution of sea ice in the north polar region. J. Geophys. Res., Vol.96, p.4791–4801.

Parkinson C.L., Cavalieri D.J. (1989). Arctic sea ice 1973–1987: seasonal, regional and interannual variability. J. Geophys. Res., Vol.C94, N10, p.14499–14524.

Parkinson C.L., Gloersen P. (1993). Global sea ice coverage. In Gurney et al. (Eds), Atlas of Satellite Observations Related to Global Change. Cambridge University Press. p.371–383.

Parkinson C.L., Comiso J.C., Zwally H.J., Cavalieri D.J., Gloersen P., Campbell W.J. (1987). Arctic sea ice, 1973–1976: satellite passive microwave observations. NASA, Washington D.C. 296 pp.

Peltier W.R. (ed.) (1993). Ice in the Climate System. (NATO ASI Series I: Global Environmental Change, Vol.12.) Springer Verlag, New York. 673 pp.

Peng T.-H., Broecker W.S., Östlund H.G. (1992). Dynamic constraints on CO_2 uptake by an iron-fertilized Antarctic. In Modeling the Earth System (Ed. by D. Ojima), Vol.3. UCAR OIES. Global Change Inst. Boulder, CO. p. 77–106.

Petropavlovskaya M.S. (1988). Remote sounding of cryolithozone. Leningrad, Gidrometeoizdat. 72 pp. (in Russian).

Pierce R.B., Grose W.L., Russell III, J.M., Tuck A.F., Swinbank R., O'Neill A. (1994). Spring dehydration in the antarctic stratospheric vortex observed by HALOE. J. Atmos. Sci., Vol.51, N20, p.2931–2941.

Polar regions and global environmental change (1993). The Globe, Issue 16, p.1,2.

Polder D., Van Santon J.H. (1946). The effective permeability of mixtures of solids. Physica, Vol.12, p.257–271.

Programme to study the climate of polar regions (1989). (Polar climatic programme of the Institute of Arctic and Antarctic.) Leningrad, Gidrometeoizdat. 82 pp. (in Russian).

Pueschel R.F., Blake D.F., Snetsinger K.G., Hansen A.D.A., Verma S., Kato K. (1992). Black carbon (soot) aerosol in the lower stratosphere and upper troposphere. Geophys. Res. Lett., Vol.19, N16, p.1659–1662.

Raatz W.E. (1984). Observations of "Arctic Haze" during the "Ptarmigan" weather reconnaissance flights, 1948–1951. Tellus, Vol.B36, N2, p.126–136.

Raatz W.E., Schnell R.C., Bodhaine B.A., Oltmans S.J. (1985). Observations of arctic haze during polar flights from Alaska to Norway. Atmos. Environ., Vol.19, N12, p.2143–2151.

Rabinovich Yu.I., Melentyev V.V. (1970). The effect of temperature and salinity on the emission of smooth water surface in the centimeter wavelength region. Trudy GGO, Issue 235, p.78–123 (in Russian).

Rabinovich Yu.I., Schchukin G.G., Novoselov A.I. (1970). The application of microwave technique for ice reconnaissance. Trudy GGO, Issue 235, p.61–71 (in Russian).

Rabinovich Yu.I., Loshchilov V.S., Shulgina E.M. (1975). Analysis of the results of measuring the ice cover characteristics (Option C). The Soviet–American Bering Sea Experiment. Proc. Final Symp. on the Results of the Soviet–American Expedition. Leningrad, Gidrometeoizdat. P.284–313 (in Russian).

Radiation and Climate (1992). Report of Fourth Session of the WCRP Working Group on

Radiative Fluxes (Palm Springs, USA, 24–27 September 1991). WCRP-69 (WMO/TD-N 471). WMO, Geneva. 47 pp.

Radionov V.F. (1994). Variability of aerosol extinction of solar radiation in Antarctica. Antarctic Science, Vol.6, N3, p.419–424.

Radionov V.F., Marshunova M.S., Rusina E.N., Lubo-Lesnichenko K.E., Pimanova Yu.E. (1994). Aerosol turbidity of the atmosphere in polar regions. Proc. RAS. Physics of the Atmosphere and Ocean, Vol.30, N6, p.797–801 (in Russian).

Radke L.F., Hobbs P.V., Bailey I.H. (1984a). Airborne observations of arctic aerosols. 3. Origins and effects of airmasses. Geophys. Res. Lett., Vol.11, N5, p.401–404.

Radke L.F., Lyons J.H., Hegg D.A., Hobbs P.V., Bailey I.H. (1984b). Airborne observations of arctic aerosols. 1. Characteristics of arctic haze. Geophys. Res. Lett., Vol.11, N5, p.393–396.

Radke L.F., Brock C.A., Lyons J.H., Hobbs P.V., Schnell R.C. (1989). Aerosol and lidar measurements of hazes in mid-latitude and polar airmasses. Atmos. Environm., Vol.23, N11, p.2417–2430.

Rahn K.A. (1985). Progress in Arctic air chemistry. Atmos. Environ., Vol.19, N12, p.1987–1994.

Raiser V.Yu., Sharkov E.A., Etkin V.S. (1975). The effect of temperature and salinity on the emission of smooth sea surface in the decimeter and meter wavelength regions. Izv. AN SSSR, FAO, Vol.11, N6, p.652–656 (in Russian).

Ramseier R.O., Gloersen P., Campbell W.J., Chang T.C. (1975). Mesoscale description for the principal Bering Sea ice experiment. Soviet–American Bering Sea Experiment. Proc. Final Symp. on the Result of the Joint Soviet–American Expedition. Leningrad, Gidrometeoizdat. p.234–270.

Randel W.J., Boville B.A., Gille J.C., Bailey P.L., Massie S.T., Kumer J.B., Mergenthaler J.L., Roche A.E. (1994). Simulation of stratospheric N_2O in the NCAR GCM2: Comparison with CLAES data and global budget analyses. J. Atmos. Sci., Vol.51, N20, p.2834–2845.

Rango A., Martinec J. (1994). Areal extent of seasonal snow cover in a changed climate. Nordic Hydrology, Vol.25, p.233–246.

Raschke E., Jacob D. (ed.) (1993). Energy Water Cycles in the Climate System (NATO ASI Ser. I: Global Environmental Change, Vol.5). Springer, Heidelberg. 467 pp.

Raschke E., Bauer P., Lutz H.J. (1992). Remote sensing of clouds and surface radiation budget over Polar Regions. Int. J. Remote Sens., Vol.13, N1, p.13–22.

Rasmussen E.A., Turner, J., Twitchell P.F. (1993). Report of a workshop on applications of new forms of satellite data in polar low research. Bull. Amer. Meteorol. Soc., Vol.74, N6, p.1057–1074.

Ray P.S. (1972). Broadband complex refraction index of ice and water. Appl. Optics, Vol.11, N8, p.1836–1844.

Report of the WMO/CAS-JSC-CCCO Meeting of Experts on the Role Sea Ice in Climate Variations (1982). (Geneva, 24–29 June, 1982). WCP-26. Geneva. 20 pp.

Report of the Seventh Session of the CAS/JSC Working Group on Numerical Experimentation (1992). (NCAR, Boneder, CO, 24–29 October 1991). WCRP-70 (WMO/TD-N 447). Geneva. 45 pp.

Report of the Fifteenth Session of the Joint Scientific Committee (1994). Geneva, Switzerland, 14–18 March. WMO/TD-N 632. 86 pp. Appendices.

Research Strategies for the US Global Change Research Program (1990). National Academy Press, Washington D.C., 291 pp.

Riedlinger S.H., Preller R.H. (1991). The development of a coupled ice–ocean model for forecasting ice conditions in the Arctic. J. Geophys. Res., Vol.96, NC9, p.16955–16978.

Robin G.de Q. (1984). Polar glaciology. NASA Techn. Memo. 86129 (part 2). Goddard Space Flight Center, Greenbelt, MD, p.37–40.

Robinson D.A., Dewey K.F. (1990). Recent secular variations in the extent of northern hemisphere snow cover. Geophys. Res. Lett., Vol.17, N10, p.1557–1560.

Robinson D.A., Serreze M.C., Barry R.G., Scharfen G., Kukla G. (1992). Large-scale patterns and variability of snowmelt and parameterized surface albedo in the Arctic basin. J. Climate, Vol.5, N10, pp.1109–1119.

Robinson D.A., Dewey K.F., Hein R.R., Jr (1993). Global snow cover monitoring: an update. Bull. Amer. Meteorol. Soc., Vol.74, N9, p.1689–1698.

Role of Polar Regions in Global Change (1990). Fairbanks, Alaska. Preprints.

Romanov V.F., Ariskina N.V., Vasilyev V.F., Lachun V.E. (1987). Atmospheric energetics in polar regions. Leningrad, Gidrometeoizdat. 296 pp. (in Russian).

Rosen J.M., Kjome N.T., Fast H., Khattatov V.U., Rudakov V.V. (1992). Penetration of Mt. Pinatubo aerosols into the north polar vortex. Geophys. Res. Lett., Vol.19, N17, p.1751–1754.

Rosier S.M., Lawrence B.N., Andrews D.G., Taylor F.W. (1994). Dynamical evolution of the northern stratosphere in early winter 1991/92, as observed by the improved stratospheric and mesospheric sounder. J. Atmos. Sci., Vol.51, N20, p.2783–2799.

Rottier P.J. (1992). Floe pair interaction event rates in the marginal ice zone. J. Geophys. Res., Vol.97, NC6, p.9390–9400.

Royer J.F., Planton S., Déqué M. (1990). A sensitivity experiment for the removal of Arctic sea ice with the French spectral general circulation model. Clim. Dynam., Vol.5, N1, p.1–18.

Russel J.M., III, Tuck A.F., Gordly L.L., Park J.H., Drayson R.S., Harries J.E., Cicerone R.S., Crutzen P.J. (1993). HALOE Antarctic observations in the spring of 1991. Geophys. Res. Lett., Vol.20, N8, p.719–722.

Rytov S.M. (1953). Theory of electrical fluctuations and thermal emission. Moscow, AN SSSR Publ. (in Russian).

Rytov S.M., Kravtsov Yu.A., Tatarsky V.I. (1978). Introduction to statistical radiophysics. Pt.2. Moscow, Nauka Publ. 464 pp. (in Russian).

Sakshang E., Slagstad D. (1992). Sea ice and wind: effects of primary productivity in the Barents Sea. Atmosphere–Ocean, Vol.30, N4, p.579–591.

Sandven S. (1992). The climatic importance of ocean and sea ice. Proc. Summer School Alpbach: Global Environment. Processes and Monitoring from Space. Austrian Space Agency. Vienna, p.64–70.

di Sarra A., Cacciani M., di Girolamo P., Fiocco G., Fua D., Knudsen B., Larsen N., Joergensen T.S. (1992). Observations of correlated behavior of stratospheric ozone and aerosol at Thule during winter 1991–1992. Geophys. Res. Lett., Vol.19, N18, p.1823–1826.

Schnell R. (1983). Arctic haze and the Arctic Gas and Aerosol Sampling Programme (AGASP). AIAA Pap. N439, 6pp.

Schnell R.C., Raatz W.E. (1984). Vertical and horizontal characteristics of Arctic haze during AGASP: Alaska Arctic. Geophys. Res. Lett., Vol.11, N5, p.369–376.

Schweiger A.J., Key J.R. (1992). Arctic cloudiness: comparison of ISCCP-C2 and Nimbus-7 satellite-derived cloud products with a surface-based climate climatology. J. Climate, Vol.5, N12, p.1514–1527.

Schweiger A.J., Serreze M.C., Key J.R. (1993). Arctic sea ice albedo: a comparison of two satellite-derived data sets. Geophys. Res. Lett., Vol.20, N1, p.41–44.

Scientific Concept of the Arctic Climate System Study (ACSYS) (1992). Report of the JSC Study Group on ACSYS. WCRP-72 (WMO/TD-N 486). 89 pp. Appendices.

Sea Ice and Climate (1988). Report of the Third Session of the JSC Working Group on Sea Ice and Climate (Oslo, Norway, 31 May–3 June 1988). WMO, Geneva. 23 pp.

Sea Ice and Climate (1990). Report of the Fourth Session of the Working Group on Sea Ice and Climate (Rome, Italy, 20–23 November 1989). WCRP-41 (WMO/TD-N 377). Geneva. 18 pp. Appendices.

Sea Ice and Climate (1992). Report of the Fifth Session of the Working Group on Sea Ice and Climate (Bremerhaven, 13–15 June 1991). WRCP/WMO, Geneva. N459, 1-II, 1–16, Annex 1–35.

Sea Ice Numerical Experimentation Group (1990). Report of the First Session (Washington D.C., 23–25 May 1989). WCRP-45 (WMO/TD-N 384). Geneva. 20 pp. Appendices.

Serreze M.C., Barry R.G. (1988). Synoptic activity in the Arctic Basin, 1979–85. J. Climate, Vol.1, N12, p.1276–1295.

Serreze M.C., Maslanik J.A., Rehder, M.C., Schnell R.C., Kahl J.D., Andreas E.L. (1992a). Theoretical heights of buoyant convection above open leads in the winter Arctic pack ice cover. J. Geophys. Res., Vol.97, NC66, p.9411–9422.

Serreze M.C., Maslanik J.A., Barry R.G., Demaria T.L. (1992b). Winter atmospheric circulation in the Arctic Basin and possible relationships to the Great Salinity Anomaly in the Northern North Atlantic. Geophys. Res. Lett., Vol.19, N3, p.293–296.

Serreze M.C., Kahl J.D., Schnell R.C. (1992c). Low-level temperature inversions of the Eurasian Arctic and comparisons with Soviet drifting station data. J. Climate, Vol.5, N6, p.615–629.

Serreze M.C., Box J.E., Barry R.G., Walsh J.E. (1993). Characteristics of arctic synoptic activity, 1952–89. J. Climate, Vol.1, N12, p.1276–1295.

Schweiger A.J., Key J. (1992). Comparison of ISCCP-C2 and Nimbus-7 satellite-derived cloud products with a surface-based cloud climatology. J. Climatology, Vol.5, N12, p.1514–1527.

Shapiro L.T. (1991). Snow on sea ice: competing effects in shaping climate. J. Geophys. Res. D, Vol.98, N9, p.17195–17208.

Shaw G.E. (1982a). Evidence for a central Eurasian source area of Arctic haze in Alaska. Nature, Vol.299, N5886, p.815–818.

Shaw G.E. (1982b). Atmospheric turbidity in the polar regions. J. Appl. Meteorol., Vol.21, N8, p.1080–1088.

Shaw G.E. (1982c). Perturbation to the atmospheric radiation field from carbonaceous aerosols. Particul. Carbon: Atmos. Life Cycle/Proc. Int. Symp., Warren, MI, October 1980. New York. p.53–68, 68–73.

Shaw G.E. (1985). On the climatic relevancy of Arctic haze: static energy balance considerations. Tellus, Vol.37B, N1, p.50–52.

Shaw G.E. (1988). Antarctic aerosols: a review, Revs. Geophys., Vol.26, N1, p.89–112.

Shaw G.E. (1993). Arctic air pollution. Proc. Russian Geographical Soc., Vol.127, N1.

Shaw G.E., Stamnes K., Hu Y.X. (1993). Arctic haze: perturbation to the radiation field. Meteorol. and Atmos. Phys., Vol.51, N3–4, p. 227–238.

Shin R.T., Kong J.A. (1982). Theory for thermal microwave emission from a homogeneous layer with rough surfaces containing spherical scatterers. J. Geophys. Res., Vol.V87, N7, p.5566–5576.

Shukla J. (ed.) (1993). Prediction of Interannual Climate Variations. Springer, Heidelberg. 265 pp.

Shulgina E.M. (1975). Radio-emission of a vertically inhomogeneous medium. Trudy GGO, Issue 331, p.64–72 (in Russian).

Shumilov O.I., Henriksen K., Raspopov O.M., Kasatkina E.A. (1992). Arctic ozone abundance and solar proton events. Geophys. Res. Lett., Vol.19, N16, p.1647–1650.

Silvente E., Legrand M. (1993). Ammonium to sulphate ratio in aerosol and snow of Greenland and Antarctic regions. Geophys. Res. Lett., Vol.20, N8, p.687–690.

Simmonds I., Budd W.F. (1991). Sensitivity of the southern hemisphere circulation to leads in the Antarctic pack ice. Quart. J. Roy. Meteorol. Soc., Vol.117, Part B, N501, p.1003–1024.

Singh R.P., Srivastav S.K., (1993). Passive-microwave radiometry of multilayer snow surfaces. Il Nuovo Cimento, Vol.16C, N4, p.437–452.

Slingo A. (1994). Clouds, cloud observations and cloud feedbacks. In Remote Sensing and Global Climate Change (ed. by R.A. Vaughan, A.P. Cracknell). Springer-Verlag, Berlin. p.269–294.

Smith S.D., Muench R.D., Pease C.H. (1990). Polynyas and leads: an overview of physical processes and environment. J. Geophys. Res., Vol.C95, N6, p.9461–9480.

Somerville R.C., Remer L.A. (1984). Cloud optical thickness feedback in the CO_2 climate problem. J. Geophys. Res., Vol.89, p.9668–9672.

Soviet–American Bering Sea Experiment (1975). Ed. by K.Ya. Kondratyev, W. Nordberg, Yu.I. Rabinovich. Leningrad, Gidrometeoizdat. 316 pp. (in Russian).

Steffen K., Abdalati W., Stroeve J. (1993). Climate sensitivity studies of the Greenland ice sheet using satellite AVHRR, SMMR, SSM/I and in situ data. Meteorol. and Atmos. Phys., Vol.51, N3–4, p.239–258.

Stössel A. (1992). Sensitivity of Southern Ocean sea ice simulation to different atmospheric forcing algorithms. Tellus, Vol.44A, p.395–413.

Stogryn A. (1984). The bilocal approximation for the affective dielectric constant of an isotropic random medium. IEEE Trans. Antennas Propagat., Vol.AP32, p.517–520.

Stonehouse B. (ed.) (1986). Arctic Air Pollution. Cambridge University Press, Cambridge. 328 pp.

Syvitski J.P.M., Andrews J.T. (1994). Climate change: numerical modelling of sedimentation and coastal processes, eastern Canadian Arctic. Arct. and Alp. Res., Vol.26, N3, p.199–212.

Takashima T., Morrissey E.G. (1976). An automated technique of determining the surface characteristics in terms of VHRR data. COSPAR Proc. Symp. on Meteorol. Observations from Space: Their Contribution to the FGGE. Philadelphia. p.145–150.

Tatarsky V.I. (1967). Wave propagation in a turbulent atmosphere. Moscow, Nauka Publ. (in Russian).

Taylor F.W., Lambert A., Grainger R.G., Rodgers C.D., Remedios J.J. (1994). Properties of northern hemisphere polar stratospheric clouds and volcanic aerosol in 1991/92 from UARS/ISAMS satellite measurements. J. Atmos. Sci., Vol.51, N20, p.3019–3026.

The Arctic Ocean Grand Challenge (1994). A decadal programme 1996–2005. Outcome of the ECOPS Euroscience Conference in Helsinki, Finland, 2–7 September 1994. International Arctic Science Committee. 67 pp.

The Ocean and Poles (1991). European Cooperation in Ocean and Polar Research. ECOPS-CES. European Science Foundation, Strasbourg. 66 pp.

Thomas R.H. (1990). Polar Research from Satellites. Joint Oceanographic Institutions, Inc., Washington D.C. 91 pp.

Thomas R.H. (1991). Polar Research from Satellites. Joint Oceanographic Institutions, Inc. Washington D.C. 91 pp.

Thomas G. and Rowntree P.R. (1992). The boreal forest and climate. Quart. J. Roy. Meteorol. Soc., Vol.118, Part B, N505, p.469–498.

Thornidike A.S. (1992). Estimates of sea ice thickness distribution using observations and theory. J. Geophys. Res., Vol.97, NC8, p.12601–12606.

Thornton D.C., Bandy A.R., Driedger A.R., III (1989). Sulfur dioxide in the North American Arctic. J. Atmos. Chem., Vol.9, N1–3, p.331–346.

Tinga W.R., Voss W.A., Blossey D.F. (1973). Generalized approach to multiphase dielectric mixture theory. J. Appl. Phys., Vol.44, p.3897–3902.

Tooma S.G. et al. (1975). Comparison of sea-ice type identification between airborne dual-frequencies microwave radiometry and standard laser/infrared techniques. J. Glaciol., Vol.15, N73, p.225–239.

Treshnikov A.F., Alexeev G.V., Mikshtas A.A., Nagurny A.P., Savchenko V.G., Khrol V.P. (1991). Ocean–atmosphere interaction in the north-polar region. Leningrad, Gidrometeoizdat. 176 pp. (in Russian).

Tsang L., Kong J.A. (1975). The brightness temperature of a half-space random medium with non-uniform temperature profile. Radio Sci., Vol.10, p.1025–1033.

Tuchkov L.T. (1968). Natural noise emissions at radiofrequencies. Moscow, Sov. Radio Publ. 161 pp. (in Russian).

Tzeng R.-Y., Bromwich D.H., Parish T.R. (1993). Present-day Antarctic climatology of the NCAR Community Model Version 1. J. Climate, Vol.6, N2, p.205–226.

Ulaby F.T., Stiles W.H. (1980). The active and passive microwave response to snow parameters. 2. Water equivalent of dry snow. J. Geophys. Res., Vol.C85, N2, p.1045–1049.

Untersteiner N. (1982). Ocean climate studies in the arctic seasonal ice zone. Large-scale oceanographic experiments in the WCRP. Papers presented at the JSC/CCCO Study Conference in Tokyo, 10–21 May 1982. WMO Publ. Ser., Vol.2, N1, p.487–510.

Untersteiner N. (1984). The cryosphere. In Global Climate, Ed. by J.T. Houghton, Cambridge University Press, Cambridge. p.121-140.

USSR/USA Bering Sea Experiment (1982). Ed. by K.Ya. Kondratyev. Amering Publ., Co., New Delhi. 307 pp.

Vinogradov V.V., Lazarenko N.N., Mironov L.V. (1975). Measurements of the sea surface radiative temperature in the northern part of the Bering Sea. Soviet–American Bering Sea Experiment. Proc. Final Symp. on Results of the Joint Soviet–American Expedition. Leningrad, Gidrometeoizdat, p.271–281.

Vinogradova A.A. (1993). Microelements in arctic aerosols (a review). Proc. RAS. Physics of the Atmos. and Ocean, Vol.29, N4, p.437-456 (in Russian).

Voeikov A.I. (1889). Snow cover, its effect on soils, climate and weather, and techniques for its survey. Proc. Russian Geogr. Soc. on general Geography, Vol.18, N2, 212 pp. (in Russian).

Vtiurina E.A. (1970). Cryogenic structure of the seasonally melting layer. Moscow: Nauka Publ. 280 pp. (in Russian).

Wadhams P. (1986). The seasonal ice zone. In N. Untersteiner (ed.), The Geophysics of Sea Ice. Plenum Press, New York. p.825–992.

Wadhams P. (1990). Evidence for thinning of the Arctic ice cover north of Greenland. Nature, Vol.345, p.795–797.

Wadhams P. (1992a). Remote sensing of snow and ice. Dundee University Summer School Seminar. 82 pp.

Wadhams P. (1992b). Sea ice thickness distribution in the Greenland Sea and Eurasian Basin. J. Geophys. Res., Vol.97, NC4, p.5331–5348.

Wadhams P. (1994a) Remote sensing of snow and ice and its relevance to climate change processes. In Remote Sensing and Global Climate Change (Ed. by R.A. Vaughan, A.P. Cracknell). Springer-Verlag, Berlin. p.303–340.

Wadhams P. (1994b). Sea ice thickness changes and their relation to climate. Geophysical Monograph 85. AGU, Washington D.C. p.337–362.

Wadhams P., Comiso J.C. (1992). The ice thickness distribution inferred using remote sensing techniques. In Microwave Remote Sensing of Sea Ice, AGU Monograph 68, Ed. by F. Carsey. American Geophysical Union, Washington D.C. p.375–383.

Wadhams P., Tucker W.B., Krabill W.B., Swift R.N., Comiso J.C., Davis N.R. (1992). Relationship between sea ice frecboard and draft in the Arctic basin, and implications for ice thickness monitoring. J. Geophys. Res., Vol.97, NC12, p.20325–20334.

Walsh J.E. (1991). The Arctic as a bellwether. Nature, Vol.352, p.19–20.

Walsh J.E., Crane R.G. (1992). A comparison of GCM simulations of Arctic climate. Geophys. Res. Lett., Vol.19, N1, p.29–32.

Walsh J.E., Johnson C.M. (1979). An analysis of Arctic sea ice fluctuations, 1953–1977. J. Phys. Oceanogr., Vol.9, p.580–591.

Walsh J.E., Lynch A., chapman W., Musgrave D. (1993). A regional model for studies of atmosphere–ice–ocean interaction in the western Arctic. Meteorol. Atmos. Phys., Vol.51, p.179–194.

Walsh J.E., Zwally H.J. (1990). Multiyear sea ice in the Arctic: model- and satellite derived. J. Geophys. Res., Vol.95, NC7, p.11613–11628.

Warren S.C. (1984). Optical constant of ice from the ultraviolet to the micro-wave. Appl. Optics., Vol.23, N8, p.1206–1225.

WCRP-41: Report of the Fourth Session of the Working Group on Sea Ice and Climate (Rome, Italy, 20–23 November 1989). WNO/TD-N 377. Geneva.

WCRP-72: Scientific Concept of the Arctic Climate System Study (ACSYS). Report of the JSC Study Group on ACSYS (Bremerhaven, Germany, 10–12 June 1991, and London, United Kingdom, 18–19 November 1991). WMO/TD-N 486. Geneva.

Weatherly J.W., Walsh J.E., Zwally H.J. (1991). Antarctic sea ice variations and seasonal air temperature relationships. J. Geophys. Res., Vol.96, p.15119–15130.

Weiner O. (1910). Zur Theorie der Refraktionkonstanten. Berichts über die verhandlungen der königlich sachsischen Gesellschaft der Wissenschaften zu Leipzig—Mathematischphysikalische Klassen, Vol.62, N5, p.256–268.

Weller G. (1992a). Arctic. In W.A. Nierenberg (ed.), Encyclopedia of Earth System Science. Academic Press, San Diego, CA. p.101–111.

Weller G. (1992b). Antarctica and the detection of environmental change. Phil. Trans. Roy. Soc. London B, Vol.338, N1285, p.201–208.

Weller G. (1993). Polar meteorology. Meteorol. and Atmos. Phys., Vol.51, N3–4, p.141–146.

Weller G., Meitner R. (1993). Satellite-derived outgoing longwave radiation, surface temperature and sea ice concentration off the coast of Adélie-Land, Eastern Antarctica. Meteorol and Atmos. Phys., Vol.51, N3–4.

Whiting G.J., Bartlett D.S., Fan S.M., Bakwin P.S., Wofsy S.C. (1992). Biosphere/atmosphere CO_2 exchange in tundra ecosystems: community characteristics and relationships with multispectral surface reflectance. J. Geophys. Res., Vol.97, ND15, p.16671–16680.

Wiesnet D.R. (1974). The role of satellite's in snow and ice measurements. NOAA Techn. Memo NESS-58. Washington D.C. 12 pp.

Wilheit T. et al. (1972). Aircraft measurements of microwave emission from arctic sea ice. Remote Sens. Environm. Vol.2, p.129–139.

Winnebrenner D.P., Bredlow J., Fung A.K., Drinkwater M.R., Nghiem S., Gow A.J., Perovich D.K., Grenfell T.C., Han H.C., Kong J.A., Lee J.K., Mudaliar S., Onstott R.G., Tsang L., West R.D. (1992). In Microwave sea ice signature modeling. In Microwave Remote Sensing of Sea Ice, AGU Monograph 68, Ed. by F. Carsey. American Geophysical Union, Washington D.C. p.137–176.

Wilkness P.E. (1989). The polar regions: research in a changing world. Bull. Amer. Meteorol. Soc., Vol.70, N2, p.160–164.

Witness. The Arctic (1994). Vol.2, N1, 15 pp.

Wofsy S.C., Sachse G.W., Gregory G.L., Blake D.R., Bradshaw J.D., Sandhoim S.T., Singh H.B., Barrick J.A., Harriss R.C., Talbot R.W., Shipham M.A., Browell E.V.,

Jacob D.J., Logan J.A. (1992). Atmospheric chemistry in the Arctic and Subarctic: influence of natural fires, industrial emissions, and stratospheric inputs. J. Geophys. Res., Vol.97, ND15, p.16731–16742.

Yang J., Neel J.D. (1993). Sea-ice interaction with the thermohaline circulation. Geophys. Res. Lett., Vol.20, N3, p.217–220.

Zakharov V.F. (1981). The arctic ice and the present natural processes. Leningrad, Gidrometeoizdat. 195 pp. (in Russian).

Zubov N.N. (1945). L'dy Arktiki. Glavsevmorputi, Moscow. 360 pp. (English translation: Arctic sea ice, Transl. 217. U.S. Nav. Hydrogr. Office, Suitland, MD, 1963. Available as AD426972 from Natl., Techn, Inf. Serv., Springfield, VA.)

Zwally H.J., Wilheit T.T., Gloersen P., Mueller J.L. (1976). Characteristics of Antarctic sea ice as determined by satelliteborne microwave images. COSPAR Proc. Symp. on Meteorol. Observ. from Space: Their Contribution to the FGGE. Philadelphia. p.94–97.

Zwally H.J., Comiso J.C., Parkinson C.L., Campbell W.J., Carsey F.D., Gloersen P. (1983). Antarctic Sea Ice, 1973–1976: Satellite Passive-Microwave Observations, NASA SP-459, National Aeronautics and Space Administration, Washington D.C. 206 pp.

Glossary of abbreviations and acronyms

AARI (AANII)	Arctic and Antarctic Research Institute
ABL	Atmospheric Boundary Layer
ACC	Antarctic Circumpolar Current
ACSYS	Arctic Climate System Study
AGCM	Atmospheric General Circulation Models
AITM	Arctic Ice Thickness Monitoring Programme
AnITM	Antarctic Ice Thickness Monitoring Programme
AOGC	Arctic Ocean Grand Challenge
APDA	Arctic Precipitation Data Archive
ARCSS	Arctic System Science
ARDB	Arctic Runoff Data Base
ARM	Atmospheric Radiation Measurement Programme
BB	black body
BESEX	Bering Sea Experiment
CAENEX	Complete Atmospheric Energetics Experiment
CART	Cloud and Radiation Testbed
CGCM	Coupled General Circulation Models
CIM	comprehensive ice map
COADS	conventional meteorological observations
DMSP	Defense Meteorological Satellite Programme
ECOPS	European Committee on Ocean and Polar Science
ENSO	El Nino/Southern Oscillations
EOS	Earth Observing System
EPICVA	European Programme of Antarctic Ice Cover
ERS	European Earth Resources Satellite
ESMR	Electrically Scanning Microwave Radiometer
FROST	First Regional Observational Experiment to Study the Troposphere
GCOS	Global Climate Observing System
GEWEX	Global Energy and Water Cycles Experiment
GFDL	Geophysical Fluid Dynamics Laboratory
GISP2	Greenland Ice Sheet Project Two

GISS	Goddard Institute for Space Studies
GOOS	Global Ocean Observing System
GTOS	Global Terrestrial Observing System
IGBP	International Geosphere-Biosphere Programme
ISCCP	International Satellites Cloud Climatology Project
INTASE	International Transantarctic Expedition
ITCZ	Intertropical Convergence Zone
ITEX	International Tundra Experiment
JERS	Japanese Earth Resources Satellite
JGOFS	Joint Global Ocean Flux Study
LATI	Land/Atmosphere/Ice Interactions
LTER	Long-Term Ecological Research
MAS	Microwave Active Spectrometer
MGO	the Voeikov Main Geophysical Observatory
MIMR	Multi-Frequency Imaging Microwave Radiometer
MRS	Meteorological Radar Station
MS	microwave spectrometer
MIZEX	Marginal Ice Zone Experiment
NCAR	National Center for Atmospheric Research
OGCM	Ocean General Circulation Models
OSU	Oregon State University
PALE	Paleoclimates of Arctic Lake and Estuaries
PELICOM	Project of Long-Term Monitoring of Sea Ice Cover Concentration
PVS	Polder–Van Santon (model)
RISE	Research of the Ross Shelf Ecosystem
RV	research vessel
SAR	Synthetic Aperture Radar
SAT	surface air temperature
SHEBA	Surface Heat Budget of the Arctic Ocean Project
SLR	Side-Looking Radar
SMMR	Scanning Multichannel Microwave Radiometer
SSMI	Special Sensor Microwave Imager
SST	sea surface temperature
TOVS	TIROS Operational Vertical Sounder
UKMO	UK Meteorological Office
VCA	video-control apparatus
WAIS	West-Antarctic Ice Sheet Studies
WCRP	World Climate Research Programme
WMO	World Meteorological Organization
WOCE	World Ocean Circulation Experiment

Index